D0759250

OXFORD STUDIES IN THE ANTHROPOLOGY OF CULTURAL FORMS

General Editors
HOWARD MORPHY AND FRED MYERS

Wrapping Culture

OXFORD STUDIES IN THE ANTHROPOLOGY OF CULTURAL FORMS

Oxford Studies in the Anthropology of Cultural Forms is a series focusing on the anthropology of art, material culture, and aesthetics. Titles in the series will concentrate on art, music, poetry, dance, ritual form, and material objects studied as components of systems of knowledge and value, as they are transformed and reproduced over time. Howard Morphy is University Lecturer in Ethnology, Oxford, and Curator at the Pitt Rivers Museum. Fred Myers is Associate Professor of Anthropology at New York University.

ALSO PUBLISHED IN THIS SERIES

Anthropology, Art, and Aesthetics
Edited by Jeremy Coote and Anthony Shelton

WRAPPING CULTURE

POLITENESS, PRESENTATION AND POWER IN JAPAN AND OTHER SOCIETIES

Joy Hendry

CLARENDON PRESS · OXFORD

Walton Street, Oxford OX2 6DP
d New York
nd Bangkok Bombay
own Dar es Salaam Delhi
Florence Hong Kong Istanbul Karachi
Kuala Lumpur Madras Madrid Melbourne
Mexico City Nairobi Paris Singapore
Taipei Tokyo Toronto

and associated companies in
Berlin Ibadan

Oxford is a trade mark of Oxford University Press

Published in the United States by
Oxford University Press Inc., New York

British Library Cataloguing in Publication Data
Data available

Library of Congress Cataloging in Publication Data
Hendry, Joy.
Wrapping culture: politeness, presentation, and power in Japan and other societies / Joy Hendry.
p. cm.—(Oxford studies in the anthropology of cultural forms)
Includes bibliographical references.
1. National characteristics, Japanese. 2. Japan—Civilization. 3. Intercultural communication—Japan. I. Title. II. Series.
DS830.H45 1993 952—dc20 92-23254
ISBN 0-19-827389-4

3 5 7 9 10 8 6 4 2

Printed in Great Britain
on acid-free paper by
Biddles Ltd
Guildford and King's Lynn

To

James and William

ACKNOWLEDGEMENTS

The research on which this book is based has been supported by many kind people in many places, over a total period of some twenty years. It is therefore impossible to mention them all here by name, although some have been acknowledged in previous publications and, to those, I owe a continuing debt. I would like to thank in particular, for the work could not proceed without them, my many friends, colleagues, and informants in Japan. For the most recent research project, I am especially indebted to Professor Teigo Yoshida and Professor Takao Suzuki, who arranged for me to be attached to the Gengo Bunka Kenkyûjo at Keio University, to Yōko Hirose, Yūji Nakanishi, Kazuko Ōnishi, Yoko Rogers (in Oxford), Takako and Minoru Shimagami, and Mrs Yonezawa, who assisted in various ways with the research, and to Jenny Davidson, who helped with domestic tasks, and with keeping my children cheerful in a foreign land.

For financial assistance, I must acknowledge the Japan Foundation, for a grant towards publication costs; the Economic and Social Research Council (United Kingdom), for their grant, reference number G0023 2254/1, which supported me during the most influential piece of research; the Nuffield Foundation for a generous Small Grant which enabled me to make a photographic expedition to Japan, as well as allowing me to pay for copies of the prints which are reproduced here; the Daiwa Foundation, which covered the costs of plates and photographs in the book; the Oxford Polytechnic Research Committee, which helped with internal expenses during the photographic expedition, which was combined with research on tattooing; Stirling University, for a travel grant which helped me to present papers on the subject in the United States; and Ian Gow, as Director of the Scottish Centre for Japanese Studies, at the same university, who created a wonderful half-time post for me so that I could devote several months of the year to writing.

Within this same university, as well as at my other place of employment, Oxford Polytechnic, 'down the road' in Oxford University, and at sundry other places of learning, I have probably driven several of my friends and colleagues nearly to distraction by going on about wrapping. I have also received support, encouragement, books, articles, and endless good examples, however, and I would like to thank for some or all of these things, Tessa Carroll, Jane Cobbi, (the late) Andrew Duff-Cooper, David Gellner, Ian Gow, Val Hamilton, Fumiko Hasegawa, Rumiko Ishii, Heita Kawakatsu, John Knight, Joseph Kyburz, Chris McDonaugh, James McMullen, Lola Martinez, Joe Moran, Rodney Needham, Jun Ohashi, Brian Powell, Ian Reader, Peter Rivière, Crispin Shore, Arthur Stockwin, Brian Street, Jim Valentine, Frank Webster, and Yōko Yamada.

Michael O'Hanlon and Howard Morphy kindly read and commented on the entire text, and Val Hamilton, Lola Martinez, and Robert Smith read substantial proportions of it. I would like to thank all five of them for many good suggestions, but, of course, take responsibility for any idiocies that remain.

Mention should also be made here of the many seminar audiences on which I have inflicted 'wrapping papers', and other parts of the book, for their comments have often inspired new lines of thinking. I would like to thank them, and, in particular, those who invited me to participate. They include Keio University, in Tokyo, where I was invited to speak to the anthropology seminar by Professors Teigo Yoshida and Hitoshi Miyake, the Linguistics Seminar at the University of Sussex (Brian Street and Ralph Grillo); anthropology department seminars at the universities of Amsterdam (Jan van Bremen); Seoul National (Kwang-ok Kim and Okpyo Moon); Columbia (Ted Bestor); the Aegean (Renee Hirschon); Oxford (Godfrey Lienhardt); Mary Washington College, Fredericksberg, Virginia (Margaret Huber), and the Anthropological Society of Hong Kong (Diana Martin); departments of Japanese Studies at Essex, Leiden (Jan van Bremen), Sheffield (Irena Powell), and the Nissan Institute, Oxford (Arthur Stockwin and Ann Waswo); the Scottish branch of the Royal Anthropological Institute (Neil Thin and Jonathan Spencer), the Royal Scottish Museum, the Japan Society of Scotland, and the Danish/Japan Society (Professor Olof Lidin and Kirsten Refsing).

I am also indebted to members of the Japan Anthropology Workshop (JAWS), because the nascent ideas which eventually developed into this book were first presented to a JAWS meeting in Jerusalem in 1987 (the paper I gave there appeared in the collection which became *Unwrapping Japan* (Eyal Ben-Ari *et al.* (eds.)), and part of Chapter 2 was presented at the JAWS meeting in Leiden in 1990 and is to appear in *Ritual in Japan* (Jan van Bremen and Lola Martinez (eds.)). I would also like to thank Carmen Blacker for inviting me to participate in the Rethinking Japanese Religion conference in Cambridge in April 1991, as this experience particularly encouraged me to investigate further ideas which appear throughout the book, although I am aware I am only scratching the surface in many cases. Some of Chapter 3 has appeared as an Occasional Paper of the Scottish Centre for Japanese Studies, University of Stirling. I am grateful for the permission of the editors to use it again.

The National Museums of Scotland have been connected in several ways with this book. First, for providing the beautiful prints, and I thank Jane Wilkinson for her help in choosing and interpreting them; for general encouragement at a low moment, I thank Jenny Calder; and at a good moment, Twin Watkins, who also enabled me to contribute to the Discovering Japan Exhibition in 1991. I would also like to thank Rupert Falkner for help at the Victoria and Albert Museum.

For help with getting the photographs together, I would like to thank Jane Bachnik, Bill Coaldrake, Bob Pomfret, Ian Reader, Peter Rivière, Yasu Takahara, Nigel Ward, the Japan Information and Cultural Centre and the Japan Travel Bureau in London, the Imperial Household Agency, and the Hutchison Library for the Disappearing World photographs. Especial thanks to Sandy Craig at Stirling University for stopping everything several times to make beautiful photocopies at various stages in the process. In the captions to the illustrations, unmarked entries indicate photographs taken by the author.

Last, but by no means least, I would like to thank Anne Goldie and Nicki Richardson in Stirling, and Edna Ackroyd in Oxford, for wonderfully supportive secretarial backup, and Tim Barton and Janet Moth at Oxford University Press for all the extra hard work the inclusion of photographs has caused.

R. J. H.

Stirling
December 1991

CONTENTS

FIGURES

PLATES

(Between pp. 112–113)

Introduction

A potential reader could be forgiven for assuming from the title of this book that the subject-matter is likely to remain in the realm of material things, that the book is about presents and how they are presented. This is true of only part of the book, however, and I hope that if the potential reader could be induced to become an actual reader, they would find by the end of the experience that they had come to alter their initial assumptions about the range of meaning of the word 'wrapping', as well as their ideas about the expression of politeness. With luck, and a spot of imagination, I hope they may even have allowed a certain blur to emerge over the line they had previously drawn between material and non-material things, though it should be emphasized at once that there is no claim made here to enter into the realms of philosophy.

The idea for the book began with an interest in the wrapping of presents, and there may well be those who regard further developments which emerge simply as metaphorical use of the vocabulary associated with such wrapping. I don't believe that the use of the word 'metaphor' in this context is particularly useful, but it really doesn't detract from the argument at this stage if that is the way it is visualized. Some of the wider uses of the notion of wrapping to be discussed here already exist in the English language, but the initial discovery of parallels in different arenas[1] of society was made in Japan with the Japanese language. Japan will thus feature as providing a model for the theory which is to be developed, but it is the aim of the book to describe principles applicable to any social system, each of which will demonstrate its own particular features.

To remain at least for the time being firmly in the English language, an idea of the scope of the endeavour may perhaps be given by a brief consideration of the notions of 'present' and 'presentation'. When we present a gift, which may also often just as well be called a 'present', we must also be concerned with how it is to be presented. This concern includes several decisions: is the gift to be wrapped, and if so, how; when and where is it to be presented, in what way, and with what form of words should the presentation be made? Are there certain clothes which should be worn in order to make the presentation properly? In other words, is it important that the person presenting the gift be presentable? Should the presentation be made in a particular place? Standing, sitting, even bowing, perhaps? And should it be made at a particular time?

Similar questions may be asked about how the present should be received, and, if it is wrapped, when and how it should be opened.

These are all aspects of gift exchange which were only indirectly considered by Marcel Mauss (1954) whose classic essay on the subject made a lasting impact in related fields. He was concerned with the meaning of gifts, but largely in the way they inevitably form part of a chain of gifts, each one carrying with it three obligations to give, to receive, and to repay. His concern is rather with the exchanges involved, with the process of exchange, and his theory may be applied to all sorts of other exchanges, both material and the less tangible. Unreciprocated dinner invitations usually dry up eventually, just as does an unreturned greeting. Mauss noted that a gift of inappropriate value may sour relations, but he did not consider the consequences of inappropriate presentation. A sumptuous home-made dinner can hardly be repaid with a stack of take-away fish and chips even if the cost is the same, and a handshake is not compatible with a low bow. Mauss did, however, note that gifts may be interpreted as a detachable part of a person, so the parallel between the presentation of the person presenting the gift, and the gift itself, could be expected.[2]

Mauss's principles of gift exchange have been adopted, applied, criticized, and adapted in many other arenas of social life, from the rather trivial examples mentioned above to complicated systems of exchange concerning trade, marriage arrangements, and other forms of long-term communication. In effect, as the work of Malinowski, Lévi-Strauss, and a number of others have demonstrated, they are among the most basic tenets of social life. I hope to show here that style of presentation is important in all these arenas, too, material and non-material. Just as there are appropriate ways to carry out transactions, there are rules about the behaviour of the parties involved, and these will vary from one society to another. Moreover, there are underlying assumptions about these activities which may rarely be made explicit, and parties who take care to investigate the assumptions of their partners in communication, may give themselves a great advantage. In some cases, if the means of presentation is elegant enough, the nature or even value of the medium of exchange may almost pale into insignificance.

At the broadest level, this book is concerned with intercultural communication, and the possibilities for misinterpretation of familiar objects in an unfamiliar context. The receipt of a nicely wrapped gift may, for example, give pleasure. It may also carry considerable further meaning and expectations it would be wise to understand. Those who are brought up in a culture where gifts are predominantly designed to give pleasure may be easily misled, easily 'bought' perhaps, just as on a personal level those who are susceptible to flattery (wrapped words) may be persuaded through their weakness. In Britain, we are rather well aware of the power of politeness, but many of us are quite bowled over by the Japanese version. My own frustration at some of the ways in which foreigners visiting Japan are charmed and ultimately outwitted in this way has been another driving force behind this book.

A supreme example of a misplaced intercultural encounter between West and East was recounted by Marshall Sahlins (1988) in a Radcliffe Brown lecture at the British Academy. He was discussing the eighteenth- and nineteenth-century British delegations which took all manner of unusual items to present to the Emperor of China as possible alternatives to their diminishing supply of silver to trade for tea. On the visit of one of the better-known emissaries, George, Lord Viscount Macartney, envoy of George III, the gifts included the latest scientific inventions, as well as all manner of gadgets designed to tempt the Chinese into creating a need for them.

They were accepted by the Emperor of the Middle Kingdom, but not for trade. He regarded them, instead, as tribute from the barbarians of distant lands. He made clear, moreover, that he had neither need nor desire for them: 'we have never valued ingenious articles, nor do we have the slightest need for your country's manufactures', he declared (Sahlins, 1988: 10). He acknowledged a 'sincere humility and obedience' (ibid.) in the accompanying missive, but regarded the 'Barbarians' tributes [as] . . . signs of the force of attraction of the imperial virtue' (ibid. 15). Macartney tried to insist on a distinction between 'presents' from another, equal sovereign, and 'tribute', and wanted to proceed to business discussions, but for the Chinese, the ceremony of presentation *was* the business of the occasion (ibid. 17).

As Sahlins points out, it was almost certainly significant that members of the British party were largely unable to understand what was going on around them, since no member of the party could converse in Chinese (ibid.), and this must have confirmed the Chinese conviction of their own superior civilization. The British were not only forced to keep finding supplies of silver, their attempt to establish diplomatic relations had also been cursorily dismissed as mere homage from the Barbarian fringes. It is by no means clear that a knowledge of Chinese language would have helped the situation at that point, but it could hardly have made things worse, and the anecdote illustrates the importance of mutual understanding as a prerequisite for successful presentations. The gift alone is evidently not enough.

Here we shall be concerned then with the significance of the wrapping of presents, with the way they are wrapped and the way they are unwrapped. We shall also be concerned with the wrapping of the people who present those presents, with the clothes they choose to wear and with other aspects of their external presentation, whether they are presenting presents or not. We shall also be concerned with the words they use, with the way they offer the presents they give, whether they be material gifts, invitations, or words of solace or consolation. We are ultimately also concerned with a much wider range of interaction, with the use of words, clothes, gifts, and any number of other devices found in the presentations people make to one another. We are concerned, also, with the relationships between these different forms of presentation in any one social context.

Erving Goffman (1959) discussed some of this subject-matter in a publication made over thirty years ago. He was concerned with the presentation of self, with the performances people put on for each other, and with the way in which they manage the impressions they create. We are concerned here too with all these aspects of presentation but, more particularly, we are concerned with culturally variable methods of presentation such as politeness, indirection, diplomacy, and dissimulation. Indeed, an initial interest in politeness preceded my interest in wrapping. In Japan, just as parcels are often composed of several layers of wrapping material, personal presentation may embody several layers of politeness. Elsewhere, a preferred mode of presentation may be frank speech, teasing, or even plain deceit. I would like to raise questions about how these modes of presentation may be reflected at other levels too.

Apart from the wrapping of gifts, the wrapping of the body, and the wrapping of language, this book will also show parallels with spatial and temporal examples of the wrapping principle it aims to elucidate. In the Japanese case, architectural style and the layout of domestic and religious edifices can be seen to use layers of 'spatial wrapping' in the way they enclose their inner sanctums, just as people use layers of 'linguistic wrapping' to express themselves. The temporal version of the wrapping principle is concerned with the ordering of meetings, drinking sessions, and other events of significance in Japanese life. To demonstrate structural principles of this sort, which underlie categories of thought and organization, may seem a trifle outmoded, but the argument does not quite end here.

Whether these things are nicely paralleled in material ways elsewhere or not, they certainly have structure and significance for the people who use them, and they may be superficially familiar to observers from outside. Problems arise when these same outsiders make assumptions about the edifices and meetings of other peoples, based largely on their own systems, and try to participate accordingly. Successful negotiations may then be a matter of luck, or pure coincidence, unless one side, at least, takes the trouble to find out about and understand the implicit assumptions of the other. Historically, Japanese people have actually been quite good at this investigation of the other, though they have managed to keep much of their own system cloaked in a veil of inscrutability.

This book will consider these various forms of wrapping in turn, illustrating their potential for communication within a particular culture, principally Japan but with examples from other cultures too, and then considering the danger of misinterpretation between cultures, particularly when the symbols are relatively familiar. The argument moves gradually to consider the way in which groups of people have impressed one another, consciously and unconsciously, in their use of these symbolic forms, and eventually turns to their potential for manipulation.

This level is to be developed, using anthropological theory of small-scale societies, into a discussion of the political and diplomatic use of language as a form of wrapping, particularly in societies which place a high value on non-verbal communication, and the suppression of direct displays of strong emotions. Language in this sense is shown to include the various material forms of wrapping already discussed, and one point eventually to be made is that members of societies which place a high value on verbal communication (and the written word) may be missing important cues in other societies which also use more subtle indicators.

The objects of this exercise are several. In the first place it makes possible a greater understanding of Japanese people and the ways in which they interact, not only with each other, but also with outsiders to their society. Armed with this understanding it should be possible for people living and working in Japan, or dealing in any way with Japanese people, to achieve smoother relations than they may hitherto have done. One of the important aims of the book is to present the arguments in a form intelligible to a non-specialist reader. The approach is based on social anthropological knowledge, but the points could well be relevant in several other fields, including and perhaps particularly, for people engaged in conducting business in an international arena.

Ultimately, however, the aim is broader than this, and I would like to use the Japanese case to demonstrate what I think may be some limitations in Western analysis of other peoples, perhaps what David Parkin has aptly called 'ethnocentric intellectual bias' (1976: 166). During my research on the subject of wrapping in its broadest sense, I have been struck repeatedly by some fundamental differences in approach to the whole notion of wrapping. These will be discussed throughout the text, starting in the very first chapter, but as a preliminary guide, I can already say that I think we are perhaps overly concerned with 'unwrapping', with revealing the perceived essence of things, where we might do well to examine a little further the nature of the concealment used.

The point is made rather well by Michael O'Hanlon in his introduction to *Reading the Skin*, a study of the significance of the elaborate bodily adornment used by the Wahgi people of the Western Highlands of New Guinea, which is a very good example of wrapping. He writes,

> In English, the notion of 'adornment' suggests the superficial, the non-essential, even the frivolous. We often think of adornment as an artificially added layer, concealing what 'really' lies beneath. The Wahgi concept is rather different. For them, the decorated appearance is more often thought to reveal than to conceal. Far from being frivolous, adornment and display are felt to be deeply implicated in politics and religion, marriage and morality. (1989: 10)

A similar point has actually been made about Japan, using the metaphor of layers of lacquer in reference to the self-restraint it is thought valuable to acquire there.

> The more coats of varnish that are laid on the foundation by laborious work throughout the years, the more valuable becomes the . . . finished product. So it is with a people . . . There is nothing spurious about it; it is not a daub to cover defects. It is at least as valuable as the substance it adorns.[3]

To choose only one current idiom, we may refer briefly to the relatively recent interest in textual analysis in anthropology. This is a prime example of the concerns of a people themselves encased in verbal modes of discourse, albeit represented in printed form, often to the virtual exclusion of pictorial illustration.[4] At one level, the works of the reflexive anthropologists (notably Geertz, 1975, 1988; Clifford and Marcus, 1986; Marcus and Fischer, 1986) are very much concerned with examining the 'wrapping' in which ethnographies are presented, the wrapping they see as put there by the ethnographers themselves. At another level, however, it is possible to discern some basic assumptions about an underlying 'essence' which they would appear to be intent on unwrapping, even if this essence is that of the anthropologists' actual relationship with the people they studied.[5]

It is the idea of the existence, or the special importance of the existence of an underlying essence which I would like to question. A recent monograph about Japan illustrates rather neatly the dilemma of the post-reflexive ethnographer desperately seeking an approach that avoids all the traps which appear to have been set. In the end, she opts for a very personal presentation, introduced in a chapter entitled 'The Eye (I)' (Kondo, 1990). Ultimately, she argues, the Japanese material reveals a Western misconception of notions of the self, which can somehow be separated from other selves with which they interact, and the book is aptly entitled *Crafting Selves*, an account of the persons with whom she, herself, interacted in Japan. How much nearer to an essence can an ethnographic account get, I wonder?

Another recent work on Japan (Edwards, 1989) bears out much of Kondo's argument about the interrelatedness of selves, in a less theoretically wrapped but no less valuable account, and it is almost certainly no coincidence that both of these writers are Americans with Japanese grandparents, seeking alongside their scholarly inquiry to identify some of the roots of their own socialization. What they both demonstrate is that ethnography, stripped of its theoretical construction, is ultimately about interaction between individuals. Interesting, certainly, but possibly not quite the essence of the people under consideration. If there is such an essence. In contrast, the recent work of a Japanese who has adopted American intellectual culture (as opposed to Japanese or half-Japanese Americans seeking Japanese culture) is so carefully embellished with contemporary rhetorical rigour that the Japaneseness of her subject-matter is almost obscured altogether (Ohnuki-Tierney, 1990).

In the end, it is almost certainly our theoretical constructs which have made the subject of social anthropology interesting, over and above the intrinsic interest most of us have in personal interaction. Perhaps we have been so concerned recently with the notions of 'deconstruction' and of 'unpacking' that we have failed to take enough notice of the construction itself, of the value of the packaging that we so quickly throw away. In the environmentally conscious green world in which we now live, perhaps I may make a plea for another level of recycling, before we find ourselves irrevocably post-.

Finally, I would like to suggest that the Japanese case presented here is merely one detailed analysis of modes of indirect communication in a particular cultural milieu, a type of social analysis which may well be applied elsewhere. The particular forms of indirection will vary from context to context but this case has proved so elegant and successful that I feel it must have wider application. It is my hope, therefore, that specialists elsewhere may find some small spark of inspiration in the idea and apply a similar process of analysis to their own material. Perhaps ultimately we may find at last that those who study Japan can contribute to the formulation of general theory about society rather than constantly falling back on its alleged unique qualities.

1

The Purpose and Meaning of Wrapping

Since gift-wrapping has proved the inspiration for the argument presented here, let us begin by examining some examples of the phenomenon. There are various reasons why a gift should be wrapped, indeed there are various reasons why anything should be wrapped, and it might be instructive first to consider some of these different reasons and how far they apply in the same way in different cultural contexts. We have already mentioned that in Japan gifts may have several layers of wrapping, and we will return to consider why this should be so, but let us first consider the act of wrapping in itself.

Mundane Wrapping

The most obvious and practical reason for wrapping goods is to protect them from outside impurities such as dirt, germs, and the vagaries of the climate. It may also be a precaution necessary to keep the contents together for the purpose of transport. Thus, a farmer or gardener with a few spare vegetables of which to dispose may bundle them into a sheet of newspaper, or an old polythene bag, to present to a visitor as a parting gesture of goodwill. This act may also protect the visitor from the 'dirt' or earth which is clinging to the roots. A supermarket may wrap a similar selection of vegetables, usually deprived of their roots, for the sake of convenience, so that shoppers can choose a pack which fits their purse, rather than waiting to have their chosen objects weighed. Certain shoppers may prefer, however, to avoid vegetables wrapped in wasteful plastic bags and choose instead to make their purchases in a market where the weighed goods are still poured into one's woven cane shopping-basket.

Properly encased chocolate bars and pre-sealed packets of boiled sweets have long been the norm, but there is still a certain appeal in having a quarter-pound weighed out from a jar, and fêtes and tourist resorts make a good deal of money selling fudge and other sticky substances unwrapped until the point of purchase. Ice-creams undoubtedly taste better when scooped into the cone from a giant tub than when plucked already encased in paper from a fridge and pressed into soggy wafers from a cardboard box. There is something unwholesome about vacuum-packed meat and cheese, though it is likely to last better than slices cut from the slab or joint, and is probably also safer than the latter from the germs of the shop assistant's breath or the flies which seem to penetrate even the best-kept kitchen. Or is this a socially conditioned view?

Within the short space of these two paragraphs, we have moved from the directly functional explanations of wrapping to a less easily classified distaste for the practice, which could well be correlated with factors peculiar to the upbringing of myself and a few of my associates who share these views. Britain (where I grew up) is not my area of expertise, and I have never carried out any systematic research here, but I suspect that there may be various elements involved in the notions expressed. Perhaps there is a certain amount of nostalgia for the past, when sticky sweets were weighed out carefully to correspond to our severely limited pocket-money; perhaps there is a sense of frustration at the waste involved in creating and almost immediately discarding all those neat little plastic bags; there may even be a serious concern with conservation. Certainly, lack of wrapping in a pre-packaged world could be seen to stand for non-standardization and individual choice.

There is more to it than this, though. The perceived distaste is also related to the way a host or hostess will carefully conceal the tin or packet from which any part of a special dinner party has emerged. It is no doubt concerned with the emphasis on home-made food as good food, with bottling and preserving, and with collecting fruit from bushes in the autumn lanes. It would seem to condemn someone to a lot of preparation, or perhaps, previously, the possession of a large retinue of servants. Fortunately for the home-makers in America, there would appear to be much more acceptance of the wrapped, canned, and partially prepared, albeit sometimes by companies with cosy sounding names[1] that almost bring the ideal mother—or comfortable slave/servant—right into the automated battery of machines which serves as a kitchen there nowadays.

We can thus already identify some cultural variation in attitudes to wrapping. We could probably do the same without even leaving the shores of the British Isles, but let us continue to look further afield. In the society in which I grew up, it was not the thing to have packets on the meal table, and during a recent family visit to New York, I discovered that my hosts (from Seattle) had been brought up with the same prohibition. For the benefit of myself and my two growing sons, they were struggling with tiny milk-jugs, until, at last, probably exasperated at the speed with which my children disposed of their milk, they reverted to their normal practice and produced the carton—right there on to the dining-table. We proceeded to compare notes about the rationales our mothers had given us for the ill-advisedness of this practice and noted how times have changed.

Plastic containers, too, are still most unacceptable in many circles in the British Isles, yet in the high-tech country which Japan has become, it can actually be a sign of status to have plastic on your table. Instead of a milk-jug and sugar-bowl, anyway certainly missing from a Japanese tea set, the way to serve (Western-style) tea and coffee in middle-class houses of Japan is with little one-helping-sized plastic cartons of cream and individually wrapped spoonfuls of sugar. The latter tend to come in long thin packets, a little different from the ones you get on airlines or in coffee shops, but the former are all but identical (Fig. 1.1)! Then, although home-made cakes and biscuits are fast catching on, it is probably still more formal to be offered a selection of bought biscuits, each enclosed in shiny decorative packets (Fig. 1.2), or an expensive delicacy from the cake shop, again often enough not only wrapped in paper, or perhaps a leaf, but its actual construction being in the form of one substance wrapped inside another (Fig. 1.3). For, as it will gradually emerge, this is a wrapping culture.

Many mundane grocery items come well wrapped in Japan, and the prestige attached to an acquaintance with Western cuisine and the ingredients it requires would almost seem to some extent to be drawn from the plastic containers deemed necessary for such goods successfully to survive a climate to which they are ill-adapted. Cooked and smoked meats are a good example, for their vacuum packs are also chosen as expensive and attractive gifts, once they are suitably arranged on a bed of satin in a substantial cardboard box (Fig. 1.4). Dairy products are another instance, cheese coming in a variety of preparations—slab, slices, even grated—but always carefully sealed in plastic, sometimes further enclosed in a tin and/or box as well. Tea bags, as if they

FIG. 1.1. Tea served with wrapped milk, sugar, and biscuits.

FIG. 1.2. Double-wrapped biscuits.

FIG. 1.3. Wrapped 'wrapped' cakes.

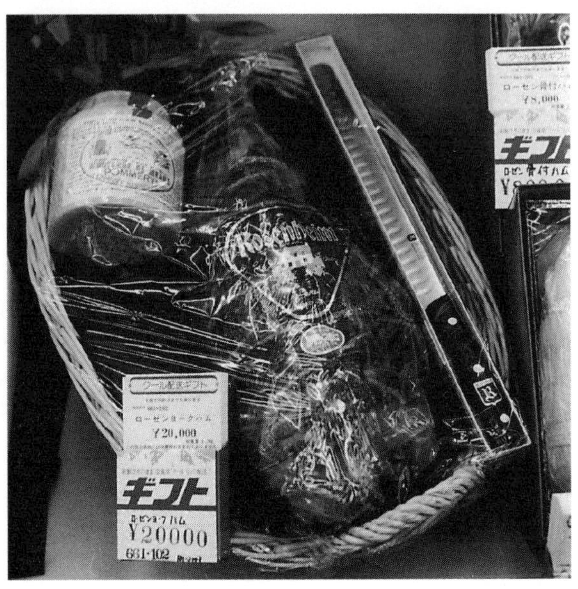

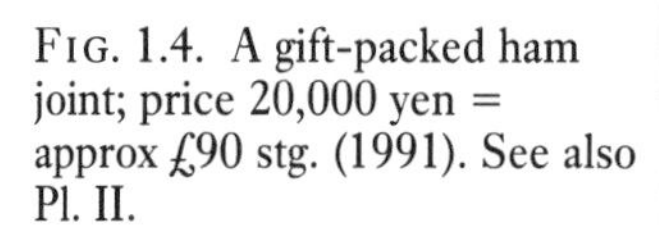
FIG. 1.4. A gift-packed ham joint; price 20,000 yen = approx £90 stg. (1991). See also Pl. II.

were not already processed enough, are usually further enclosed in plastic packets of five before being packed in the cellophane-sealed box which contains them elsewhere too.

Strange combinations of cultural ideas emerge in the international scenes created in airline cuisine. My own home company—the only one to have granted me the opportunity to sample all their classes—demonstrates clearly the aversion for plastic I have already described as one moves higher up the social (or should I say economic) scale. In the business section, the totally plastic suite of containers characteristic of economy class has undergone modifications, such as the introduction of glassware and metal cutlery, whereas those who can afford the luxurious surroundings of the first-class cabin are served all their meals from silver platters on to china plates. On the trip between London and Tokyo, the sudden appearance of a selection of 'high-class' wrapped Japanese delicacies, somewhat wilted from the airline conditions, is something of a cultural shock to the native passengers, although the Japanese guests for whom they are presumably designed may have a totally different view of things.

Gift-Wrapping

After this somewhat lengthy preamble, let us move on to consider some of the reasons why gifts should be wrapped. The paper is expensive, often used only once, and in Britain it is usually ripped off soon after the present has been handed over, then only to be fairly speedily discarded.[2] Yet, the donor may have spent some considerable time making its selection, choosing a matching ribbon, and folding and encasing the object into a shape as attractive as his or her skills would allow. In some cases, an extra sum may even have been paid to have the gift wrapped professionally. At Christmas the resulting object may be granted a period of display under the tree, and some admiring remarks may be made about the appearance of a wrapped gift on any occasion, but this is a transient piece of art, and in this country the focus of interest is really on the object inside the wrapping.

It is no doubt for this reason that it is regarded as polite to open the gift in front of the donor, and then make suitably happy noises about the contents, whether these noises express the true impression made by the present or not. Here, then, we have reached a reason for wrapping—to introduce an element of surprise. Many of us prefer, at least in anticipation, to be given something we neither need nor want, if it is a surprise, than to inform a relative about what we would like to be given in advance of our birthday. Of course, the art of true caring and friendship is to know what will please one's loved ones so that the surprise will also be a genuine pleasure, but we all also have experiences of dissimulation in these circumstances, and the phrase 'it's the thought that counts' is well worn as a charitable excuse of failure in this respect.

This is not the end of the explanation of wrapping, however, as we also wrap gifts which have been requested, even when the recipient is well aware of the contents. It is a custom to hand over gifts enclosed in fancy paper, and if the object is too large we may just make a token wrapping of a part or two of it. The paper itself may carry a clear indication of the occasion which it marks—a 'happy birthday', a Santa in a sleigh, a baby and a stork, or a set of wedding-bells. The gift may express all kinds of things about the relationship between the people involved, as Mauss (1954) has so ably discussed, but the wrapping is a chance to mark the occasion in an appropriate way, to introduce a festive air, a sense of ritual. It is simply the proper way to present a gift.

Wrapping also provides an opportunity for individuals to express their taste in choice of paper, and their economic success in its quality. Some donors like to choose the most expensive paper available, perhaps so that the recipient will feel worth the extra cost. Others like to make their own wrappings out of odds and ends, and an English book devoted to the subject of gift-wrapping claims to help the reader avoid the cost of expensive paper with its 'over a hundred bright and clever ideas for wrapping gifts inexpensively and with creative flair' (Burdett, 1987: 7). Again, the special effort put into the decoration of the gift is probably expected to transmit a message of goodwill to the recipient, although it could also say something about the proclivity of the donor to engage in artistically creative activity.

All this interpretation is fine within its cultural context, but customs and understandings do not always find their parallels in other places where they are used. Even within the countries of Europe there are a number of variations in wrapping custom. In some it is usual for stores to wrap gifts as a service when they are purchased, the paper then displaying to the recipient the name and prestige of the supplier. In others, it is considered an art to be accomplished at designing and creating one's own presents from as imaginative a selection of materials as one can muster. The rules of presentation and receipt also vary, as do the exclamations it is thought appropriate to utter. In the United States, it is the practice on occasions such as birthdays and showers to gather all the presents and their donors together so that the opening can be carried out as a kind of ceremony.

In Japan, however, although this American custom has been imported for children's parties, it was previously more usual to put a gift on one side until the donor had departed. The immediate opening of a gift is said to display too much interest in the material content of the offering rather than in the sentiment it expresses. It is even quite common for people in receipt of many gifts to keep them in their wrapped splendour, once they have ascertained their value, and then redistribute them when the occasion arises. Evidently here we are in the realm of quite a different set of values. Surprise hardly figures, and the wrapping is in fact said to be more concerned with purity and the careful separation of the recipient from any pollution the donor may carry (M. Araki, 1978: 19). It is also often more important to choose an object of an appropriate value than one the recipient will treasure.[3] In recent times, with Western influence, the 'personal' gift has come to have a place between intimates in Japanese society, but it is also less important that a gift to an intimate friend or relative be properly wrapped.

It is already evident that the language of gift exchange and the wrapping involved varies from one culture to another, and if one is to avoid offence, or at least misunderstanding, in another system, one must learn to recognize different signals embodied in familiar materials. Several writers have discussed Japanese gift exchange (e.g. Cobbi, 1988; Morsbach, 1977) and Harumi Befu (1966, 1968) has written about the specific case of gift exchange between Japanese and Americans and the possible disasters and even breakdown of relations which can arise as each side struggles to outdo the other.[4] Anthropologists have long been aware of the importance of understanding different views of the world, and they are also aware of the problems associated with their own cultural biases. It is my view that as we work more and more in the industrialized, cosmopolitan world, with its reciprocal influences, we must be particularly careful to avoid assumptions about and therefore misinterpretation of customs which seem similar to our own (cf. Cheal, 1988: 3–4).

This need to reassign value to familiar symbols applies not only to gift exchange but to many other areas of intercultural communication. Indeed, it applies to all the areas which will be discussed in subsequent chapters under the various guises of 'wrapping'. It is often also a very difficult task, and if it is taxing for anthropologists, whose bread and butter it forms, the need for the task itself is often only dimly recognized by others to whom it could make a vital difference to their everyday lives. To press this point firmly home at the outset, let us first describe in more detail some further aspects of Japanese wrapping.

Japanese Wrapping—Practical Details

For most exchanges of gifts in Japan, as elsewhere, there are customary ways of wrapping and presenting the article in question. There are also occasions when it is appropriate to present gifts, and there are commodities which are thought suitable to be presented at those times. Books of etiquette advise about details of wrapping practice (e.g. M. Araki, 1978; Ogasawara, 1985; Watanabe, 1989) but in modern Japan it is often possible to have a gift wrapped in the appropriate way at the point of purchase, and most stores will include the price of the wrapping paper and any decoration in the cost of the gift. Some stores will offer a range of wrapping materials from which a choice may be made, others will charge extra for coverings which differ from the regular wrap they provide. One specialist shop in Tokyo has even gone to the trouble of publishing its own hardbacked manual, with English versions of the explanations (G'Area Communication Project, 1988). The custom of wrapping gifts purchased in a store was probably imported from America, and a decorative bow of the same origin is often used for gifts made on imported occasions such as birthdays and Christmas, but indigenous occasions have more complicated rules.

First of all, there is a clear division into two types amongst gifts made in the more traditional Japanese order, these being gifts for happy or auspicious occasions associated with celebrations of life, and gifts for memorials for the dead. The paper used for the former is decorated with an emblem known as a *noshi*[5] (Fig. 1.5, top right), properly a small piece of abalone wrapped in a hexagonally shaped open envelope, but often simply printed on the paper itself. This piece of shellfish officially indicates to the recipient that the donor of the gift is free from the pollution associated with death, when it would be prohibited to use meat of any kind, although it has recently simply become a customary symbol of a gift. Paper for memorial gifts is decorated with a suitable motif such as lotus flowers (Fig. 1.6). Further distinctions are made in the way of folding the paper, a more ancient indicator of meaning with up to some 500 possibilities (Ogasawara, 1985: 42–3), and if the gifts are to be tied up, the strings are in different colours. Nowadays printed paper will often have these colours depicted across them instead of string.

In fact the string colours may indicate further divisions in types of gifts.

According to Ekiguchi, this *mizuhiki*[6] (Fig. 1.5), as it is called in Japanese, comes in eight different combinations of colours. For auspicious occasions, crimson and white represents the most formality; there may also be red and white, red and gold, or silver and gold, and a multicoloured combination represents an informal occasion. Black and white, blue and white, or just plain white are used for funerals and condolence or memorial gifts. There are also various different ways of tying these strings, some representing more or less formality, others having more specific significance, such as a knot which cannot be untied and which therefore symbolizes that it is for an event, like a wedding or a funeral, which should not be repeated.

On any gift, whether it is tied with *mizuhiki* or not, there is usually a space above the real or printed knot to write the specific purpose of the gift, and below, to write the name of the donor. Money is often presented as a gift in Japan, and in this case, it is enclosed in specially made envelopes with similar distinctions to those described above (Fig. 1.7). There is also a space on the back of the envelope to write in the amount. It is possible to buy a range of such envelopes with a variety of bows and decorations already affixed, and certain adornments are suitable for certain occasions, just as the more elaborate envelopes would be inappropriate for a small denomination of banknotes inside.

An additional aspect of the wrapping of money is interesting and less easy to interpret. This is the custom of wrapping the money in a plain white sheet of paper before it is placed in the envelope (Fig. 1.8). This layer of wrapping may be compared to the way a material gift is usually completely enclosed in another layer of paper before the *noshi* or memorial paper is placed around it. In some cases gifts may be found to be wrapped several times, some of the layers having obvious functions like protection or postage, others apparently being purely decorative (Figs. 1.6 and 1.9). This idea may be exemplified by describing the kind of gifts which are presented in midsummer, generally by people who are in some way indebted to the recipients.

The content of the packages is often enough a domestic commodity such as food or drink, although perhaps imported and therefore slightly luxurious in nature, and it will be first of all encased in the usual way, in bottles, cans, jars, or whatever. The set of goods will then be carefully arranged, typically on a bed of satin, in a presentation box, or possibly a basket, and sealed into place with cellophane (Fig. 1.10). On purchase, such a display will be further wrapped, have an appropriate *noshi* layer fixed around it, and the whole parcel will then like as not be placed in a carrier bag with the name of the store printed on the outside. Custom requires that the gift be handed over in this form, so the items inside may actually be encased some six times.

It would seem that the number of layers of wrapping is in some way indicative of the formality of the occasion, although according to one opinion (Oka, 1988), a single sheet is used to denote an event such as a funeral that

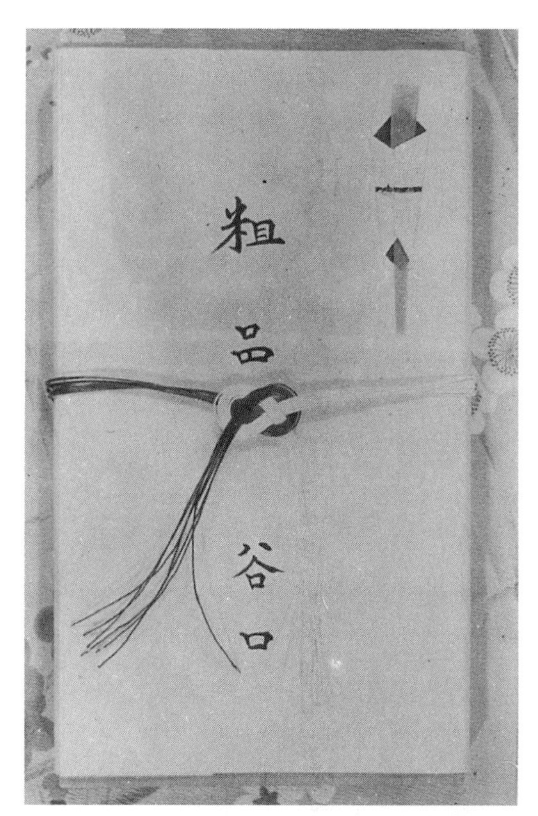

FIG. 1.5. A formally wrapped gift with *noshi* and crimson and white *mizuhiki*; courtesy Oxford Polytechnic. See also Pl. III.

FIG. 1.6. A condolence gift, with lotus motif.

Fig. 1.7. Envelopes for wrapping money may also have a *noshi*, if for a happy occasion, and *mizuhiki*. See also Pls. IV and V.

Fig. 1.8. Money may be wrapped in an inner packet as well as the decorative one; these envelopes are for condolence, the left one with a Buddhist expression, the right a Christian one; courtesy Bob Pomfret.

should happen only once, so that it may also be a silent confirmation of an auspicious occasion which could bear repeating (signalling a message similar to our 'many happy returns'), but no one has mentioned this specifically. Instead, indigenous explanations talk in terms of the 'care' being expressed, care for the object and therefore care for the person to whom it is being presented (Uno, 1985: 118–19; cf. M. Araki, 1978: 20). The Japanese word *teinei* used in this context can be translated both as 'careful' and as 'polite', so that expressing care for others in this way is synonymous with polite behaviour in a Japanese view.

It is also thought to be polite to wrap a single-page letter in a plain sheet of writing-paper, although it would apparently be even better to write more than

FIG. 1.9. Well-wrapped cakes.

one sheet, and I have received several more missives expressing this form of politeness. For example, when my son's passport had to be left behind at the travel agent's office for a correction to be made to a visa, it arrived a few days later in an envelope big enough to hold a foolscap folder. This envelope opened, I drew out a slightly smaller one, sealed again but with nothing written upon it. Opening this one, I found yet another smaller sealed envelope inside, again with nothing written on the outside. My son's passport was to be found inside. On another occasion I received some photographs by post. The number of envelopes was the same as with the passport, but the photographs were then further sealed inside a plastic packet inside the smallest envelope.

Another expression of the value of wrapping in a Japanese view is to be found in the way precious objects are stored. Unlike many Western collectors, who display the objects of their interest in a glass case, or in some other visible location, the owners of precious objects in Japan are quite likely to keep their things well hidden away. If a visitor should express an interest, he or she will be treated to an often deliberately slow and careful unpacking process. Pots, for example, may be folded first in silk or some other soft material, then placed in a purpose-built box, itself often a work of some art, and the box may even be wrapped again in paper or some other substance (Fig. 1.11).

Indeed, some artefacts may be enclosed in multiple boxes. The Kizaemon tea bowl, described and discussed in an essay by the folk-artist Sōetsu Yanagi, is apparently considered to be the finest in the world. Yanagi had been

FIG. 1.10. *Chūgen*, midsummer presentation melon.

wanting for years to see the bowl, and some element of suspense is communicated in Bernard Leach's English adaptation of his work. 'It was within box after box, five deep, buried in wool and wrapped in purple silk' (Yanagi, 1972: 191). One afficionado of ceramics told me that a collector's best pots are kept for only a very few eyes, since there appears to be some idea that every unwrapping is detracting a little from the value of the item. Thus a visitor may gauge his status in a collector's house by the number of pots he is shown.

This reluctance to expose precious objects to the air may have some practical basis in the case of ancient scrolls or garments, which may be longer preserved in this way. Scrolls often spend most of the year packed away, appearing only to be hung for brief periods as the season or a particular occasion determines. Kimonos and other Japanese garments, too, are usually kept wrapped up in stout paper parcels except on the rare occasions when they are to be worn. They are thus protected from the ravages of the Japanese climate and the attack of the local insect population. Nevertheless, there is more to this custom than these practicalities. For objects which would suffer not at all from exposure to the outside world are given the same treatment, and there seems to be some value attached to the wrapped state itself.

According to Robert Smith, an art object without a box may well lose most of its value (personal communication). The boxes are often made specially for the object, and will bear the name of the pot on the lid and the signature and

FIG. 1.11. A precious pot is carefully wrapped and kept in a box; by courtesy of the Board of Trustees of the Victoria & Albert Museum.

seal of the artist on the reverse. He recounts the horror of a friend, who took up a position in a large North American museum, when she discovered that the 'packing' in which Japanese ceramics had arrived had been disposed of in the interest of saving storage space. 'As long as they don't ever want to sell any of them,' she is quoted as having commented.

Japanese Wrapping—Some Indigenous Views

Japanese are very proud of their wrapping skills and they have for some time been exporting their ideas about wrapping and packaging to other parts of the world. A book which made a great impression, partly inspired by an exhibit requested for an international exhibition on packaging by the Museum of Modern Art in New York, is Hideyuki Oka's splendid 1967 collection known as *How to Wrap Five Eggs*. This sumptuous coffee-table book contains an abundance of photographs taken by the author, mostly of traditional, time-consuming packaging using natural materials such as straw and dried leaves. The collection was first put together for an exhibition in Japan and the book was originally published in Japanese, but the success of the English version is evident in that a subsequent volume, *How to Wrap Five More Eggs*, was published in 1975. The exhibition, which started life in Japan in 1963, moved from the Japan House Gallery in New York in 1972 to appear in nearly a hundred locations across three continents.

Another glossy book with numerous detailed diagrams and coloured illustrations was published in 1985, and in paperback in 1986, entitled: *Gift-Wrapping: Creative Ideas from Japan*. Its author Kunio Ekiguchi explains in the introduction that 'the concept of wrapping, *tsutsumi*, is not limited to the function of packaging. It plays a central role in a wide variety of spiritual and cultural aspects of Japanese life' (1986: 6). He reiterates the element of care involved in noting that just as one helps a friend into a coat carefully and courteously, a gift should be wrapped tenderly and conscientiously. 'In Japan,' he asserts, 'it is said that giving a gift is like wrapping one's heart' (ibid. 7).[7]

Similar sentiments are expressed in the catalogue of the exhibition entitled 'The Art of Japanese Packages' which toured various museums in Canada in 1988, and here the aesthetic aspects of the creations are emphasized as well. The author of the introduction to the catalogue, again Hideyuki Oka (1988), takes the line that traditional forms of packaging are dying out in this modern industrialized age, and he laments their 'slipping back into the mists of history'. He sees this as a serious loss of human love in a world where taking the time and trouble to create a beautifully wrapped object is seen as 'inefficient and unproductive'.

> Even though the object to be wrapped may be no more than a small confection, someone who truly wants to please, who wants it to taste even more delicious, will go to great trouble to wrap it carefully by hand . . . This is something the human hand must do, and only loving care enables us to perform such troublesome manual tasks to the very end.

He also notes that as packaging techniques reached the highest level of perfection, they truly became an art form, so that 'packaging transcended its simple role as something in which to wrap certain items and became an end in itself'.

Ekiguchi, although he admits that many complex forms of wrapping have fallen into disuse, is perhaps less pessimistic than Oka and writes in the present tense about the continuing importance of wrapping in today's Japan. He also mentions the idea that wrapping and the care taken with it reflect the sentiments of the giver, and he notes that 'the Japanese have always considered it discourteous simply to pass an unwrapped, unconcealed object from one hand to another' (1986: 6). He sees this as part of the restraint which he argues has become synonymous with refinement in Japan, and he explains that wrapping, particularly in white paper, symbolizes that the gift is free of contamination and impurities (ibid.).

This point is mentioned also by Oka, who talks of the sacred aspect of isolating the clean from the unclean, an activity characteristic of the ritual of the indigenous Shinto, a religion very concerned with purity and pollution.

Wrapping something in white paper, which itself has religious connotations (see Chapter 2), separates it from the dirt and pollution of the outside world, and, according to Araki Makio, purges the heart of the presenter from sin, so that no bad feeling will be transferred to the recipient (1978: 19; cf. Ogasawara, 1985).[8]

A final example of the export of Japanese wrapping ideas is to be found in another coffee-table volume, *Package Design in Japan* (1989), which appears in English, French, and German. It includes examples of packaging created by a number of Japanese designers, together with a collection of their views on the subject. These range from the animistic idea that Japanese packaging is made to house the 'being' which is present in any object (ibid. 8), through the more practical notion that packaging is, for designers, a tool of communication (ibid. 13), to the unwritten rule that 'however expensive the contents may be, the sender must express through the wrapping that he considers the present a small one' (ibid. 44).

The influence of this export of wrapping ideas from Japan is also finding its way into Western thoughts on the subject, as Julia Foster tells her readers in a book entitled *Presents*. 'The concept of improving the beauty of an object through its wrapping is most profound in Japanese culture and is central to their traditional sense of beauty,' she writes, but she goes on to demonstrate the ease of projecting one's own culturally biased expectations on to the customs of another people. 'The Japanese concept that something should be beautifully concealed, no matter how troublesome or inefficient the act may be, so that whoever receives it will actively enjoy opening the present' (1986: 90) suggests the element of surprise which we value, whereas total concealment is not necessarily always the aim of Japanese wrapping, nor is the enjoyment of the act of opening.

The works which have so far appeared in English to explain Japanese wrapping methods provide only a fragmentary glimpse of the possible interpretations of this phenomenon, and they also tend to be rather idealistic. In practice in everyday life in modern Japan an enormous number of beautifully wrapped gifts change hands for a large variety of different ostensible reasons, but the wrapping process may have very little connection with the notion of human love. As Mauss (1954) was at pains to demonstrate, gifts which are in theory voluntary, are often subject to strict bonds of obligation, and Japan is no exception to this rule. Cobbi (1988: 113) makes an important point when she notes that the hedonistic perspective—'une notion de plaisir'—associated with the French word *cadeau* is not generally the same in the case of Japanese gifts, and elsewhere there may even be entirely inauspicious associations (e.g. Parry, 1986; Raheja, 1988). People wrap gifts, or, more often in today's world of department stores, have them wrapped, because this is the appropriate way to present them. Without wrapping, the gift would fail to carry the message as properly intended, and the procurement and delivery of the gifts one is obliged to present may indeed be a very time-consuming and troublesome activity.

It is true that a gift wrapped by hand in home-made paper may in certain circles carry a special message of love in today's Japan, as it may have in times past, but a gift entirely unwrapped may carry the same message. Indeed, many of my Japanese informants explained to me that the formal wrapping of a gift expresses a certain distance in a relationship, so to leave a present unwrapped is a way of expressing intimacy. Without a context it is of course difficult to make a complete interpretation of the message transmitted by a gift, and the authors above were limited severely in the space they had to write about the subject, but the notion of wrapping as a measure of refinement in Japanese society would suggest that people in using it are constrained by motives beyond those of human sentiment.

In recent years there has been considerable concern, in the press, and amongst consumers themselves, about the waste involved in all these layers of refinement (e.g. *Asahi*, 18 May 1990; *Yomiuri*, 8 January 1991). Surveys usually report a majority of respondents in favour of cutting down on this extravagant use of world resources (e.g. *Japan Times Weekly*, 22 July 1991: 21), and the Ministry of Trade and Industry has created a committee to encourage limits (*Focus Japan*, March 1991: 3). Some individuals have taken to recycling paper with their own paper-making machines as a move towards conservation. This is a good solution, for the Japanese propensity to wrap would seem to touch on much deeper underlying thought patterns, as I hope will be shown in the chapters which follow.

The logograph used to write the concept of wrapping in Japanese is quite descriptive, easily seen as two lines enveloping one another 包. An interesting variation of this character is portrayed at the beginning of a videotape about Japanese wrapping, where the character suddenly becomes a mother holding a child (Fig. 1.12). A colloquial word used by Japanese men in reference to their mothers can be literally translated as 'bag' (*fukuro*), although it usually carries an honorific *o* in front of it. This notion of mother as a kind of wrapping for her child, even after it is born and grown up, has in fact been shown by psychologist Yōko Yamada to be a dominant image amongst Japanese respondents to a questionnaire she devised to examine how the mother–child relationship is perceived by students.

She asked the students to make drawings of the images they have of their relationship with their mothers, first a memory of when they were small, and then currently while studying at university. The answers included many symbolic depictions which could be interpreted as wrapping, some developing through time, but still involving an idea of wrapping. Yamada used the image as the title of her book on the subject—*Watashi o tsutsumu haha naru mono* (1988), which translates approximately as 'mother who wraps me'. It is of course difficult to say whether this image is a peculiarly Japanese one, but

FIG. 1.12. The character for 'wrapping' turns into a mother with child at the beginning of a film on the subject.

Yamada has subsequently carried out a similar study in the United Kingdom so it will be interesting to see her results.

There is certainly a notion of power associated with an enclosed state, like that of a child in the womb, in Japanese folklore, which includes stories about supernatural children born from a peach, a gourd, and a segment of bamboo. The idea of a 'power of seclusion' is associated with the practice, known as *komori*, of ascetics who shut themselves away to gain religious enlightenment (Blacker, 1975: 98). The power-giving qualities of enclosed vessels such as these are encapsulated in a concept of *utsubo*, a word discussed years ago by the folklorist Origuchi as having several applications, but always implying the notion of being wrapped up: *mono ni tsutsumarete iru koto* (1945: 267).

Origuchi's idea is that containment in a vessel of some kind nurtures spiritual power. Origuchi used this idea to interpret the mysterious *shinza*, an unused bed-seat which forms part of the Japanese enthronement ceremony, to suggest that originally the emperor would have lain here, 'wrapped like a cocoon in the coverlet', ready to emerge fully empowered to take on the role of the new emperor (Blacker, 1990: 191). This idea would fit well with the common symbolism of death and rebirth associated with such rites of passage (van Gennep, 1960), although it is by no means universally accepted by

Japanese commentators (ibid.). The concept of *utsubo* nevertheless provides another example of Japanese views of wrapping.

Iwao Nukada, the author of a comprehensive book in Japanese (1977) devoted to the subject of wrapping, goes into much more detail about the variety and meaning of wrapping in Japan. He discusses the way in which wrapping has developed aesthetic, religious, and magical qualities over and above the original functional ones, as well as becoming subject to strict rules of etiquette and courtesy. He too couches these developments in the context of a scheme of cultural evolution or civilization (something akin to that of Norbert Elias, 1978), providing a plethora of examples from different historical periods to support his view, which dovetails with the notion of value placed on refinement and restraint outlined by Ekiguchi. To be able to wrap things properly becomes a measure of refinement and civilization which is applied within Japanese society, but also by implication since Japan is supposed to be so advanced in this respect, as a way of giving Japan an edge over the rest of the world in cultural achievement.

Nukada further divides types of wrapping into three: the wrapping of goods, which has been the focus of this chapter, the wrapping of the body, and the wrapping of space. In subsequent chapters we will be considering Nukada's second and third types, but we will also turn to examine the notion of wrapping in a wider context which includes language as a form of wrapping. On the whole his theory about refinement applies there, too. It is a phenomenon by no means peculiar to Japan, and it is in considering this type of wrapping that we will find ourselves most comfortably in an area which lends itself to cross-cultural comparison.

2

The Language and Value of Japanese Wrapping

It would appear from the differences we have briefly outlined in the possible purposes of wrapping that the Japanese case at least offers great scope for putting meaning into the activity, and we may even be mistaken to try and attach too much importance to the actual gift inside at all. Perhaps it is rather in the wrapping that we should seek most of the significance of a Japanese gift, although the monetary value of the enclosed object should be of an appropriate amount, as is indicated below. A Western perception of the practice prepares us to regard wrapping as a means to obscure the object inside, whereas in a Japanese view it would seem that the function of wrapping is rather to refine the object, to add to it layers of meaning which it could not carry in its unwrapped form.

There are, of course, considerations of the reciprocity and ongoing exchange of which the gift may form part. In fact, in view of the number of foreigners who find it very difficult to maintain any form of reciprocity with gift exchange in Japan, it may be necessary to revise our ideas on this score too. This would not necessarily be to deny Maussian theory, however, as there may well be non-tangible benefits moving at least in one direction. In this chapter we will develop this argument about the value attached to wrapping itself, by considering some particular customs associated with the presentation of gifts and examining the role of wrapping in these specific cases.

It will become clear that these two aspects of presentation—the object enclosed and the wrapping enclosing it—cannot in fact be separated from each other, so that the total meaning of any presentation must include the importance and value of the wrapping used. In the first instance we discuss there is even some ambiguity as to whether the gift itself has merely been reduced to wrapping, and the significance of this case in the cycle of reciprocity is considered. The second example is concerned with the value of a particular type of wrapping material, itself sometimes presented as a gift.

In the second part of the chapter we turn to look at the wider value of wrapping materials when they are used in other ways. This approach contributes to the general value of wrapping in a Japanese view, and it eventually allows the discussion of wrapping to broaden out so that language may be considered as valuable in this respect too. This brings us to the point at which the material model of wrapping may be applied in other arenas. In general, the purpose of this chapter is to demonstrate the variety of interpretation possible in the presentation of gifts of one sort or another, the importance this mode of communication may have in a particular culture, and the possible dangers of misunderstanding at an intercultural level.

The Currency of Towels and Sheets

There is some variety in the type of commodity which is given as a gift in Japan, but there are also occasions when the object appropriate to be presented remains fairly constant and, furthermore, the only wrapping required is a transparent cellophane bag or a fold of paper which conceals only part of the goods. A good example is the use of small towels as presents. It is a widespread practice for shops and other commercial ventures to present towels to their customers, perhaps regularly, as at New Year, or simply in recognition of a substantial purchase. The towel is much like any other towel except for the fact that it will have the shop's name and address clearly marked upon it. Hotels are also inclined to have a supply of such towels, particularly if they offer hot spring water in their baths, although they are increasingly likely to charge for them rather than handing them out as gifts. Either way, the towel is taken home and probably used and washed frequently, the advantage for the business concerned being that of reminding the recipient of the location where it was acquired. It can thus serve as a form of advertising while perhaps touching a chord of nostalgia at the same time (Fig. 2.1).

That this practice is not simply a form of advertising, however, can be seen by comparison with another common custom, namely that of presenting towels to one's new neighbours. In this case, the towels have no name and address on them, but simply offer an opportunity for a family to initiate relations with their new neighbours. The usual procedure is to present oneself at the front door of each of the closest neighbours in turn (strictly speaking, the 'three opposite and the one on either side', although the closest group may be defined rather differently), to announce that one is new in the neighbourhood, and, while handing over the towel, to make an unspecified, formalized request for benevolence (*yoroshiku onegai shimasu*). There are one or two other possible gifts used at such a time (see e.g. Bestor, 1989: 194–5), but in reference to the practice, informants often describe the items as *taoru toka* (a towel or something like it). In both these cases the hope is expressed that the relationship signified by the movement of the towel will continue to exist.

It is probably useful to mention at this stage that the towels used at bath time in Japan are much less variable than those used elsewhere, both in size and shape. Their dimensions are around 76–80 cm. long and 29–32 cm. wide, barely larger than a hand towel in the British order of things, and the same towel is used both for washing and, once rung out, for drying. Some

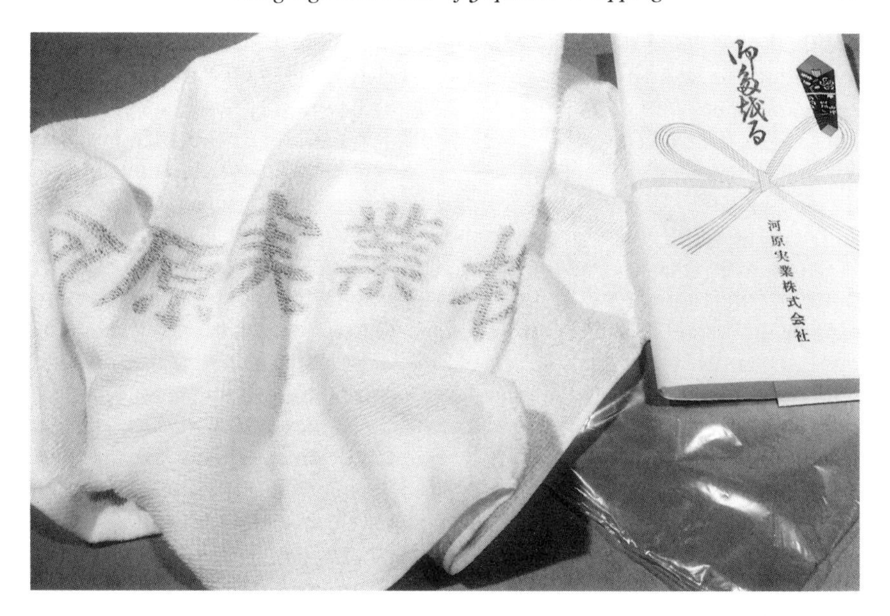

FIG. 2.1. A presentation towel, with its wrapping; courtesy Bob Pomfret.

FIG. 2.2. Towel animals; courtesy Bob Pomfret.

people do have and use larger towels, but usually in addition to, rather than instead of, the standard ones, and huge towels termed 'bath towels' in English are sometimes even found used as bedding, as if they were sheets. Most households have a very plentiful supply of the small standard towels, probably partly because they are received so often as gifts, and they are doubled up to be used as floor cloths once their initial freshness has worn off.

There is usually also a good supply of sheets, whether in the form of Western bath towels or not, in most houses, and apart from their primary function, they are used in a rather similar way to towels, which may help to explain the degree of conceptual overlap. This time they are not given away quite so lightly, but, without a great deal of wrapping, may express a closer relationship than towels. At a funeral, for example, those who attend present a sum of money to the family of the deceased, and they usually receive a gift as a sort of token return. The sum of money depends on the proximity of relationship as does the gift received. Distant acquaintances are usually offered a handkerchief, but the closest relatives may well be given sheets.

An event which brings some of these household gifts into a single category is a school or kindergarten bazaar. This is the occasion for turning out things which have accumulated in one's house, and many 'gifts' find their way on to the local equivalent of the white elephant stall. Handkerchiefs, towels, and sheets are usually available in abundance, at prices considerably reduced from those found in stores, and these are accompanied by other common gifts such as sets of bowls or glasses. None have been used, but their packaging has lost the pristine quality which makes them presentable as gifts, so their value is now merely domestic, evidently much less in that form than it was when they made a round as gifts! And this time the money paid for them goes into school funds so they have acquired an extra value on the way.

An interesting variation on the theme was something I observed and experienced in an association with the Cub Scouts of Japan, to which one of my sons belonged. The mothers of a particular pack had been asked to hold a stall at a bazaar, and we gathered to make a few items to brighten up the usual arrangement of worn-out gifts. The raw material was none other than an assortment of towels, probably in most cases received as gifts. We pulled, twisted, and sewed them up, attached a few bows and buttons, and almost before we knew it we had created a collection of charming little animals (Fig. 2.2). These were sold at a price not much different from the original cost of the unadulterated towel, albeit in its presentable form, but they fetched more than they would have in their battered cellophane bags, and their creation provided an occasion for charitable socializing amongst the mothers involved.

The point being made here is that the towels and sheets in question have become a kind of currency[1] in the way they are being used. These objects have very clear functions in a domestic situation, but this main function is only a small part of the story of their lives. Eventually, they may be used by someone for their ostensible purpose, but they may also first have been on a round of presentations, if one person was able to pass them on to another before the packaging became too creased. In this way, they can be compared with money, which is also often presented as a gift in Japan, but their use is limited to certain appropriate occasions and they would probably also represent too small a sum to be presented as cash. The school bazaar provides an occasion for redistribution since some families will have accumulated more of these objects than others, depending on the nature of their social and economic relationships.

The towels and sheets do not require a great number of layers of wrapping, for much of their meaning as presentations is embodied in the object itself. Their intrinsic value is very little, their functional value only very domestic and personal. It is almost as if they are regarded as objects without any monetary value at all, and the significance of this would be clearly that their receipt calls for no reciprocal gift, they do not need to be repaid. Here the importance of Mauss's second obligation is seen. If one is presented with a gift, one is under an obligation to receive it, on pain of breaking off or rejecting the relationship. However, the option is, at least theoretically, there, and it is quite possible to return a gift, unopened, in order to express a rejection of the relationship it implies. It would thus be inappropriate, and somewhat risky, to make a substantial presentation to someone with whom one has as yet no relationship, for it immediately burdens that person with decisions about whether or not to accept it. An insubstantial gift, such as a towel, is perhaps merely an overture, its acceptance little more than a polite gesture, so that its lack of value is entirely appropriate in this case.

It would be inappropriate then for the contents of the bag to be obscured since it would not immediately be clear that this was such an insubstantial presentation, rather than a gift proper. Usually, the wrapping of a gift anyway provides some sort of indication about its contents (cf. Morsbach, 1977: 105–6), and in this case, the lack of opaque wrapping paper has its own significance. In the end, the towel can perhaps be seen as a present and its wrapping all rolled into one, as long as it is encased in some way, and it looks new and clean. The meaning is there in the fact that it has changed hands, and further meaning may be printed on to it by the donor. It is not meant to elicit a return gift, simply to express a relationship, which the donor hopes will continue. In the terms used by David Cheal, such gifts express the 'availability' of the parties in a relationship to each other (1988: 18). The recipient is obliged only to acknowledge this relationship, although those presenting the towels may well be hoping for more.

The role of the towel here is plainly to communicate at a level beyond that of words, expressions which may actually be quite hard to put into words, or that it would be uncomfortable to be more explicit about. In Japan high value is placed on non-verbal communication of one sort or another, and it would be easy for those from a more explicit culture to miss some of the cues. Non-verbal communication is of course by no means limited to Japan, indeed, there are several other areas around the Pacific rim where it is, as in Japan, very bad form to express oneself directly, particularly on negative subjects. It is perhaps no coincidence then that many of Mauss's examples in *The Gift* came from this very ethnographic region.

The work of Malinowski (1972) on the Trobriand Islanders, for example, made very clear how much value the ceremonial presentation of gifts has for the people as a whole, as well as for the expression of differences of status within specific groups. A more recent study of the same people by Annette Weiner (1983) has revealed even more information about the symbolic value of objects in that society. Here it is quite taboo to mention directly any kind of negative feelings one may have about other people, and the anthropologist herself recounts with some clarity an occasion of near disaster when she began to upbraid a local person whom she felt had damaged her bicycle. The individual left immediately, and her mistake was only clarified in subsequent discussions with other people, when she learned that a more suitable moment to express smouldering dissatisfaction with someone is by the speed and efficiency—or, rather, by the lack of it—used when helping them to pick yams. An annual occurrence, when neighbours turn out in force to help one another, is also the occasion for measuring one's popularity or otherwise, apparently, and this is plain for all to see in the size of the pile of yams one is able to accumulate.

This is of course an extreme case, and in everyday life people express satisfaction and dissatisfaction with each other in various ways, very often using material objects as vehicles for doing this. Any number of small symbolic gestures may be required to ensure the mutual goodwill of parties engaged in a social relationship, and their mode of presentation may offer a means of qualifying the meaning of the symbol.

Furoshiki—*Towel or Wrapping?*

In support of the argument that the towel may be regarded simultaneously as a present and its wrapping, it might be useful to consider another form of Japanese wrapping not yet mentioned, namely the *furoshiki.* This is a piece of cloth, usually square in shape, and found in a variety of sizes. It is used predominantly to carry things about, first having secured the object inside by tying opposite corners of the cloth together to form a handle which may be grasped. *Furoshiki* may be fairly plain, but they also come in a variety of beautiful patterns and decorations, so that they may be given themselves as gifts (Fig. 2.3). Their use has lapsed somewhat in recent years, with the introduction of a huge number of purpose-built bags and cases, but *furoshiki*

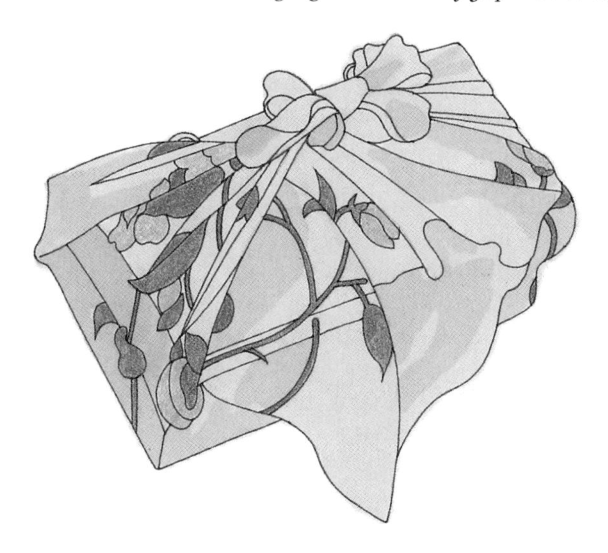

FIG. 2.3. A parcel wrapped in a *furoshiki*; drawing by Lyn North, copyright Japan Library. See also Pl. VI.

are still to be seen on the streets, and, in my experience in the country, they are used particularly to envelop and carry a present from one house to another. This serves the dual purpose of containing the item in a way convenient to transport, and concealing its precise contents from the eyes of curious passers-by. On arrival at its destination, the *furoshiki* may be removed, or, for many anticipated occasions, the present may be handed over in the *furoshiki*, which would then be used to wrap some small token return gift as an expression of appreciation.

An even more interesting comparison may be made here in view of the fact that a *furoshiki* was previously used as a towel. According to Yōko Yamada (1989), this object was originally a combination of a bath towel, a bath mat, and a clothes basket. It would be used, first, to carry clean clothes to the bath house, secondly as a bath towel, thirdly as a bath mat, and finally, to wrap and carry the soiled clothes home again. In this way it is an excellent example of the flexibility for which it has also been praised even only for its adaptable carrying capacities (Ii, 1989). Apart from its practical functions, however, the *furoshiki* also seems to have had some important ritual functions in its own right, which add further layers of meaning to Japanese forms of wrapping.

In Kyushu, for example, the *furoshiki* has a special form known as a *minofuroshiki*, reported until recently at least in islands of Nagasaki prefecture (Tanakamaru, 1987). This *furoshiki* seems to have formed an important part of a woman's trousseau, used again at several ceremonial occasions throughout her life. Consisting of three pieces of cloth sewn together, it would bear the crest and the name of the woman's house of birth. It was presented after the wedding, symbolizing the link being set up between the families. The *minofuroshiki* was used to cover the boxes to be sent with the bride to her new home, it was used to carry presents back to her natal family when she returned to visit, and it was kept carefully amongst a woman's most valuable possessions. In some poorer families, the bride's trousseau would be sent in parts over the years after the union was initially set up, and the *minofuroshiki* would cover the last load, symbolizing the bride's complete separation from her former home. After this had arrived, the marriage was formally cemented, and the bride was treated with new respect since until that time there was always a possibility that the arrangement might break down.

The *minofuroshiki* was thus a status symbol, standing for the security of the bride in her new home, but also for the link she retained with her former family. It was laid on the woman's coffin when she died, and passed on down through the family, thus keeping alive for several generations the memory of the family ties created by that particular marriage. Similar traditions seem to have been found in the prefectures of Saga and Fukuoka, as well as Nagasaki (Tanakamaru, 1987), although the *furoshiki* was there known simply as a *yomeiri-furoshiki*, an item still reported until the present day in the Karatsu district of Saga-ken, where old women are said to bring them into hospital with them, and handle them tenderly as if for comfort. These *furoshiki* have been described as a combination of a *furoshiki* and a *noshi* (ibid.), indicating also the status and position of their owners, and certainly serving a very clear symbolic function in representing links between individuals and households.

According to the Korean commentator, O'Young Lee (known as Orion Ii in Japanese) the use of the *furoshiki* is increasing again in recent years. He has written a book about what he calls '*furoshiki* culture', in which he contrasts flexible aspects of Eastern lifestyle with the more fitted ones of the West (Ii, 1989). For some time examples of the latter—such as suitcases—represented modernity as they were introduced from the West, but he argues that in the post-modern world the advantages of the *furoshiki*-style of life are becoming appreciated again, in the West as well as the East. One example he gives for this argument is related to the use of *furoshiki* in Japan, and its equivalent the *pojagi* in Korea, for carrying babies around. These methods are contrasted with pushing prams, and since the practice of carrying has recently spread to the Western world, Lee argues that this is one way in which the old-fashioned characteristics of *furoshiki* culture are being revived.

According to Lee, too, it was common practice in Japan to print patterns and the household *mon* on to *furoshiki*, from the time, in the Muromachi period, of the shogun Yoshimitsu, who apparently introduced this idea to avoid confusion at bath houses (Ii, 1989: 22). It was a short step, then, to use them for advertising, which is indeed what happened. Daimaru department store was apparently the first to profit from the idea by printing the distinctive characters of their name on to the *furoshiki* which enveloped their goods and thus displayed their name all the way down the Tokaido route from Edo (Tokyo) to Kyoto. The carrier bag has undoubtedly effectively taken over this function for the large department stores, and we thus find an interesting predecessor for this now ubiquitous outer layer of wrapping.

Lee's argument continues to develop the apparent Eastern preference for flexibility, discussing in some detail the use of screens instead of walls, chopsticks instead of knives, forks and spoons, and futons and *zabuton* (cushions) rather than beds and chairs. With clothes, he contrasts the tidy fit of Western tailoring with the looser, more flexible cut of Korean and Japanese garments, which leave room for expansion or shrinkage. At the level of language, too, he argues that some things should be left deliberately unclear, and talking too much closes off the possibilities for flexibility. In fact he, as a Korean, chides the Japanese for what he sees as too much concession to Western logic, which he argues leaves no room for nuance, discussion or argument. Language should not be clear-cut, he argues (Ii, 1989).

Lee's thesis serves to emphasize two important points being made in this book more generally. The first is concerned with the value placed on leaving language open to interpretation beyond the spoken word, which Lee regards as an example of flexibility. I would like to suggest that this leads also to a greater use of other means of communication, such as through material objects, than in cultures where language is more specific. Secondly, the way in which Lee moves from discussing *furoshiki* to illustrating his argument in other areas such as language, clothing, and the use of space also nicely serves as a model for the whole organization of this book, too, as will be seen in subsequent chapters.

A recent ethnographic account of the culture of the Yekuana people of Venezuela, entitled *To Weave and Sing*, illustrates implicitly the importance of this wider view of culture. The author describes at the beginning of the book how he set out to obtain a translation of a creation epic known as *Watunna*, said to belong to these people, but discovered it was impossible to pin this knowledge down to a narrative account. In some frustration at first, he began to apply himself to making the beautiful baskets used by these people, and as he went through all the practical and ritual procedures of collecting and treating the materials, deciding on and creating their designs, he began to realize how much more he had learned about *Watunna* than by simply asking questions. He writes: 'To understand the *Watunna* I had originally come to learn demanded much more than just verbal skills. It required the use of all my senses or, more precisely, a reorientation to the nature of meaning and the manner of its transmission' (Guss, 1989: 4).

Omiyage—*Souvenirs*

By way of contrast to the above discussion about towels, another type of gift which is commonly presented in Japan offers great scope for variety in its content, indeed novelty value is probably its chief merit, although its wrapping is also of paramount importance. This is a gift brought back by travellers to those who have remained at home. In this case, the gift often does represent a form of reciprocity, for there is also a custom whereby those who are staying at home present travellers with a gift, usually of money, on their departure. It is apparently usual for half this sum to be spent on a return gift for the donor (Graburn, 1983: 45), although travellers will probably bring something back for all their close relatives and associates regardless of whether they have received such a sum. They will, however, have a clear idea how much must be spent on each person, according to their proximity in social terms, and they will choose gifts accordingly.

In this case the precise content of the gift is less important than its value in monetary terms, but the shops to be found in tourist resorts, stations, and airports always have a range of any particular foodstuffs or other gifts in different quantities and therefore at different prices (Fig. 2.4). The gift must be a local speciality, preferably something unavailable in one's home region, and the characters used to write the Japanese name for such a gift, namely *omiyage*, literally mean 'local product'. The luggage racks on trains, buses, and planes usually sport abundant evidence of this practice, for the gifts must be properly wrapped in paper and carrier bags which make clear that the enclosed objects have been brought from the place in question. A passenger returning home very often carries at least as much volume of souvenir gifts as regular luggage (Fig. 2.5).

This practice goes back for some hundreds of years (Kyburz, 1988), although it is obviously dependent on the ability to travel. In the pre-modern period, travel for pleasure was less likely, at least ostensibly, than travel for religious purposes, and a widespread custom was for a group of people to save money to send one or more representatives on an annual pilgrimage.[2] In the end each member was expected to get a chance to do the travelling, but in the interim all the members could have a small share in the experience when they had contributed to the expenses by receiving a gift from the area concerned. Kyburz argues that the parting gift and return gift taken together form a 'magical' link destined to bring the traveller safely home. It also brought some of the religious benefits to the members of the group who remained behind.

In this respect the *omiyage* are like *omamori* and *ofuda*, amulets or talismans which are regarded as carrying some of the power of a divinity, to be channelled into protection or aid. They are purchased at shrines and temples to this day, again often during journeys, and they may have specific purposes such as road safety, good luck with exams, health, and prosperity in everyday life. They have been described as 'conduits through which sacred power... flows' (Swanger, 1981: 237), although Kyburz emphasizes the symbolic association—a rake to bring in prosperity, a beckoning cat to draw in customers (1991: 108–9). Either way the power lies less in the object itself

FIG. 2.4. Souvenirs (*omiyage*) come with a variety of different prices.

FIG. 2.5. Souvenirs may double the volume of luggage.

than in ideas about its power to transmit divine assistance, and, indeed, after a certain period, both types of object become useless and should be destroyed.

Omamori and *ofuda* are very often wrapped, too (Fig. 2.6). Indeed, they may again consist of little more than wrapping, although they usually carry words—the name of a temple or shrine, part of a sutra—and some are thought to lose their power if they become unwrapped (McVeigh, 1991: 151–6). They may be made of paper, wood, or cloth, or layers of one enclosed in layers of the other. In *omamori* it may be hard to distinguish between the wrapping and the wrapped, but it doesn't really matter. If they have been purchased in a shrine or temple associated with the appropriate power, and they are treated in an appropriate way, they can be expected to have an appropriate effect.

With *omiyage* it would not really do to buy such a present just as one arrived home, for the meaning of the gift lies in the very fact that it has been brought from a particular place, possibly far away.[3] The local wrapping paper makes clear that this is indeed the case, whatever the contents. Kyburz comments, however, that nowadays travellers have replaced gifts with religious meaning by products with famous brand names, such as Burberry scarves from England,[4] which possibly adds a strong economic element of meaning to the gift, since these are expensive items. Famous products such as these are also known as *meisan*, specialities of a particular area, and as such they carry some of their meaning intrinsically. In fact, it is perfectly possible to buy Burberry items and a large number of other brand-name products in Japan, and

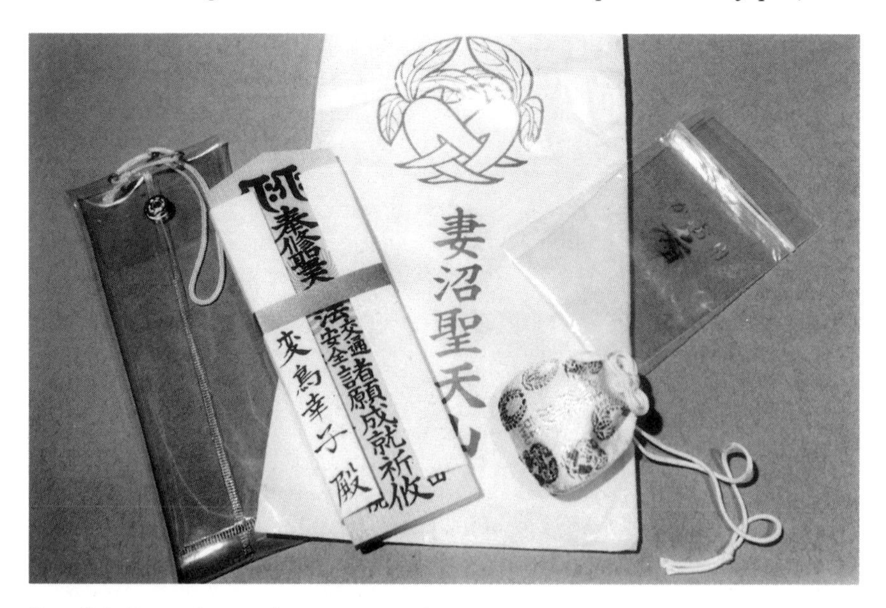

FIG. 2.6. Protective amulets are wrapped, too.

airports and stations around the world help forgetful travellers to make up for lost opportunities.

In the past, too, gifts from particular areas were of course carried by travellers all over the world to present to those they visited. Sometimes these took the form of a kind of tribute, as when local Japanese lords returned to the capital after visiting their home area. At other times, these gifts had political or diplomatic significance, perhaps implying requests to pass through a territory, or an opening of trade relations. Presentation was of course important here too, and a classic example of failure to convey the appropriate meaning, or anyway rejection of that meaning if understood, was the case discussed in the Introduction of eighteenth- and nineteenth-century British delegations to the Emperor of China, when presents intended to stimulate an interest in trade were merely accepted, even rather ungraciously, as tribute.

This anecdote illustrates the importance of mutual understanding as a prerequisite for successful presentations. The gift alone is not enough. It will represent a relation of some sort, but the way it is presented, possibly its wrapping or other details of decoration, will also carry meaning, and if this is misunderstood, the whole purpose of the presentation can be lost. This point is particularly important in the case of international or intercultural exchange, but the wider point being made applies to any case of presentation. The meaning must be understood by both sides, and it is unlikely to happen automatically in the case of intercultural transactions. In the particular case of Japanese presentations, much of the meaning is to be found in the layers of wrapping.

The word *miyage* is nowadays applied to gifts other than those brought from a long distance away, indeed it has almost come to take on a meaning as wide as 'present'. It refers, for example, to gifts taken when visiting a friend's house, whether just for a cup of tea or coffee, or a meal, or even to stay for a period. It refers, of course, to gifts taken from one person travelling from one area to visit someone in another, and in this case, it may carry some of its literal associations mentioned above, especially if local produce is chosen as the content of the gift. Otherwise, wrapping may well indicate something about the relationship between the people, close friends using little more than that necessary to enclose the object or objects, more distant relations preparing a package suitable to the formality of the occasion.

This is not the kind of gift that can be handed on with its original meaning, but it is certainly not uncommon for people to share, for some other purpose, presents received in this way. The wrapped cakes and biscuits served to impress visitors, discussed in the last chapter, is one possible way, and such an offering would also provide an opportunity to indicate the range of one's acquaintances and their travels. Gifts received in this way may also be shared with neighbours or other friends who drop in, perhaps by way of expressing thanks for a small favour; indeed, the receipt of gifts from afar actually provides an excuse to share the produce and may in this way present a suitable opportunity for discharging some kind of obligation, tangible or otherwise. In these cases, the content of the gift is not very important, although the knowledge that someone likes a particular product would increase its aptness for sharing. Most of its value in fact lies in the wrapping which identifies it as an *omiyage* suitable for redistribution.

The Value of Wrapping Materials

In general, the materials for wrapping are valued in other ways in Japanese usage, and in this section a consideration of some of these will bolster the argument about the general value of wrapping over and above its functional element.

PAPER

The most common element used for wrapping in modern Japan is paper, a commodity which is easy to fold and crease, and on which a variety of motifs and patterns can be imprinted. Paper is indeed used in many other countries to wrap objects and indicate their nature on the outside, and some varieties of Japanese wrapping paper have certainly received foreign influence, but the most highly prized wrapping paper in Japan is plain white *washi*, literally 'Japanese paper'. To a Western eye this looks a coarse, unrefined creation, but this is part of its quality. Paper itself is accorded special significance in several other ways in Japan.

First of all, it is used in ritual, particularly that associated with Shinto activities. Before almost any ceremony, the priest must prepare various utensils and many of these involve the careful cutting of paper into appropriate shapes. A sacred space is marked off with rope and paper streamers (Fig. 2.7), for example, and branches of the *sakaki* tree are decorated with paper to be offered by participants at the altar. A part of any Shinto ceremony is a rite of purification, carried out when the priest shakes a staff over the participants.

This staff, known in Japanese as a *gohei*, again comprises a stick decorated with paper streamers. According to the Japanese folklorist Yanagita (1971: 263), who has investigated the development of the object, it can be interpreted as a *yorishiro*, a Japanese word which describes a place to receive the divine, a vehicle of spiritual power. Less orthodox religious practitioners also make use of paper in their creation of ritual paraphernalia, although sometimes this is gold or silver, rather than just plain white. One man of my acquaintance solemnly shakes a few drops of sake over a gold-paper 'tree' every evening in order to bring good fortune to his house.

FIG. 2.7. Paper marks off a sacred area.

Nukada (1977: 139) argues that the importance of paper in indigenous Japanese religious activities is related to the fact that the Japanese word for paper (*kami*) is homophonous with the word for god (*kami*), although the two take different Chinese characters for their readings. Thus, in a Shinto view, a piece of white paper signifies the purity of the gods and it is used for the purpose of purification. Whether Nukada's argument is correct or not, there is certainly a notion of purity associated with white paper, which separates anything it wraps from the impurities of the outside world, and brings it to a recipient free of pollution. Carefully made folds demonstrate that the paper has only been used once, and these are said to act as a kind of seal (M. Araki, 1978: 19–20).

During several historical periods it has been customary for a daughter of the emperor to perform a special ritual function at the main imperial shrine at Ise, assiduously practising wrapping, folding, and tying as a means to know and understand the will of the gods (Nukada, 1977: 134–8). This practice is undoubtedly related to the form of the talismans and amulets, which were described in the previous section, as consisting of little more than appropriately folded paper, though these are usually wrapping a prayer or some other object such as a tiny statue. The more substantial *fuda*, which is placed in the household shrine, may have the prayer written on a tablet of wood or stiff card, but these are again invariably wrapped in paper and usually tied up neatly as well.

Paper has great aesthetic qualities in a Japanese view, too, perhaps the best-known example being the paper doors and windows which make up many of the partitions in a Japanese house. Notable and fairly standard is the soft light reflected and refracted by the *shōji*, sliding windows with thin tissue-like paper stretched where glass would be found in a Western version. Offering more variety is the thicker and altogether more durable paper used to decorate the *fusuma*, or sliding doors, which divide houses into rooms and serve as doors for the ubiquitous large cupboards built into Japanese houses (Fig. 2.9). These may be plain white, again, or sport designs, perhaps a mountainous scene, depending on the fashion of the period. In both cases, the paper is renewed from time to time, just as in Western houses walls are decorated and redecorated.

Expensive paper is purchased for calligraphy and the creation of hanging scrolls. In an exhibition of writing or painting, commentators will examine the paper as well as the work upon it, and sometimes notes are provided about the paper which has been selected. It is often thought in the West that Japanese paper is made exclusively from rice, but a paper-maker in the community in Kyushu where I worked made very high quality calligraphy paper out of bamboo shavings he collected regularly from a chopstick factory. The style and skill of the writer or painter is of course important in Japanese art, but special care is taken to select paper suitable for an overall effect.

Another almost universal use of paper in Japan is in the practice known as *origami*, literally 'paper-folding'. Children learn this art at kindergarten and school, and packets of *origami* paper make popular gifts from adults to children, often providing a good opportunity for co-operative fun as the pair exchange knowledge about the shapes they can create. There may be a ritual function here, too, as when people work to create a string of 1,000 paper cranes in order to ask for some kind of divine assistance, commonly to aid the healing process when somebody is ill. More substantial gifts can be created using similar principles, and some quite elaborate dolls and other objects may be created entirely from folded paper (Fig. 2.8). These activities have also had religious significance in previous times (Ogasawara, 1985: 42).

STRAW

Straw used to be another important material employed in wrapping although it has been replaced somewhat recently by synthetic materials with comparable qualities. It is still found in the wrapping of bottles, eggs, and for the presentation of fish under some special ritual circumstances. There is, for example, a custom found in Kyushu where a newly married couple visit the bride's parents on New Year's Eve with a large yellowtail (*buri*) fish which is served to the family's guests over the New Year period (Hendry, 1981*a*: 189). As Oka (1967) points out, straw's structure of hollow tubes allows it to provide ventilation for perishable goods and, if carefully used, it can help to preserve

FIG. 2.8. An elaborate paper doll displayed in the Paper Museum in Tokyo.

them. Its stiff yet pliable qualities also allow considerable scope for adornment and decoration with a veritable galaxy of variety.

Perhaps for similar reasons, straw has also been used for many ritual purposes. Clearly visible all over Japan, for example, are the large straw plaited ropes, or *shimenawa*, which hang in the archways at the entrance to shrines (Fig. 2.10). These serve to mark off the sacred space inside, and smaller versions of them may be used to give ritual value to other objects. One example is the adornment with straw of the *tai*, or sea bream, which often forms part of a bride's betrothal gifts. A thin straw rope, decorated with paper as mentioned above, is used to mark off sacred space for the ceremony preceding the construction of private houses and many public buildings, and in preparation for festivals in some parts of the country such a rope is hung around the whole district. Straw is also the raw material for creating objects

FIG. 2.9. A Japanese room uses paper windows, paper sliding doors, and floors of straw matting, to achieve its aesthetic value; courtesy Japan Information and Cultural Centre.

such as octopus, fish, and sake cups which are hung out over rivers at an annual river festival held in many parts of the country after the rice has been transplanted.

Straw is, of course, used in most countries where it is a by-product of crops grown, for the creation of objects both functional and decorative (see e.g. Sandford, 1974). In Britain, corn dollies are a good example, and these have also been associated with ritual activities and imbued with religious power, although now commonly described as 'pagan'. The growing of crops has always been a vital activity fraught with possible dangers, and it is hardly surprising that material so flexible and so readily available should be used in ritual ways. Its use may well, therefore, reflect important cognitive categories in the people concerned, and in the Japanese case, it is not surprising that it has been used for various forms of wrapping. Its additional use for the marking off of the sacred would seem to give an extra value to the practice of wrapping, especially in view of the time and care invested in straw wrappings, and of the similar association already described for paper.

FIG. 2.10. A straw *shimenawa* marks off a sacred area.

WOOD

Many beautiful containers are made out of the abundant wood which is to be found in the forests of the Japanese archipelago. Paulownia, Japanese cedar, and cypress are amongst the varieties used, and an important element of the construction is the way in which the wood is cut. *Masame* is the term given to a highly prized method involving wood cut across the grain from the centre of the trunk and split by hand to reveal a very fine straight grain. Boxes created from this wood will not bend or shrink, and gifts presented in such a container are apparently a sign of great respect (Oka, 1988). In general, the value of an object will be reflected in the way it is encased, and precious objects are usually accorded fairly sturdy boxes, sometimes almost *objets d'art* in their own right. A gift of highly prized foodstuffs may be enclosed in a box, too, and this has the effect of adding a layer of politeness and/or respect to the meaning of the presentation.

As with paper, care is taken to choose beautiful wood for the interior of Japanese houses, and its qualities are also prized for their aesthetic effect. A special trunk is usually chosen for the principal post, known as the *daikokubashira* (a term also used as a metaphor for the head of the house), which stands at the inside end of the *tokonoma* (see Chapter 5), and older houses will have splendid polished beams in the ceiling. Some also have decorative wooden scenes above the *fusuma*, and most have rooms with entirely wooden floors. The wood used in shrines and temples is also carefully chosen for its aesthetic qualities. It is not unknown, for example, for a master carpenter to go into the mountain forest and choose his materials before they are cut down (Coaldrake, 1990: 20).

Wood is used in ritual ways, too, and an interesting indigenous example is the Ainu *inau* (Fig. 2.11), a ritual staff which closely resembles the *gohei* used for purification purposes in a Shinto rite. In this case the staff is usually created from natural wood, shaved in such a way that long curls hang down, like the paper streamers of the *gohei*. The *inau* was highly valued by the Ainu, being regarded again as a vehicle of spiritual power (Munro, 1962: 28–9). According to Munro, the staff is a modified form of objects which used to be stylized effigies of human beings, made this way to be receptacles for ancestral communication.

Throughout Japan, the tubs and barrels which serve as sake containers for betrothal gifts and other ceremonial occasions also have ritual value. The use

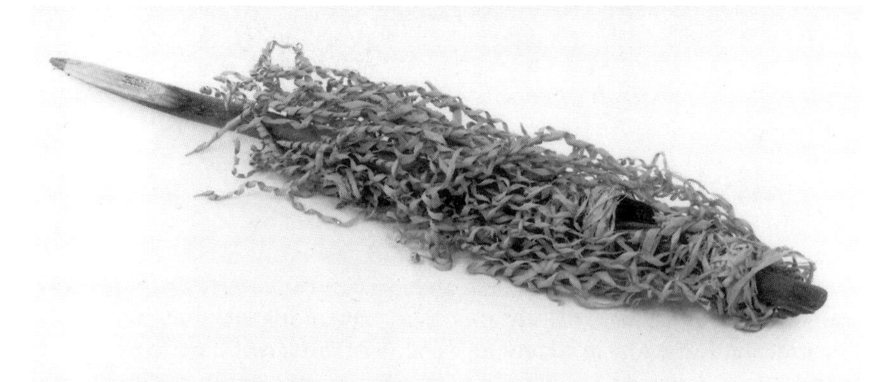

FIG. 2.11. The Ainu *inau* was regarded as a vehicle of spiritual power; © The Trustees of the National Museums of Scotland 1993.

FIG. 2.12. Sake served from a red tub into a wooden box ritualizes an occasion; © The Trustees of the National Museums of Scotland 1993. See also Pl. VII.

of fragrant wooden containers for the drinking of sake on such occasions would seem to add a special value to the sake, as well as helping to ritualize its consumption (Fig. 2.12). Sake-drinking forms a part of many ceremonial occasions in Japan, and a nest of three red lacquered cups of different sizes, thereby fitting inside each other, is the usual set of utensils for the principal rite involved (Fig. 2.13). The crux of a wedding, for example, is an exchange referred to as a 'three-three-nine times' (*sansankudo*) after the way the bride and groom take turns to drink between them three times out of each of the three cups. The same set of cups is brought out to celebrate the naming of a baby, or the passing of adults into retirement.

The sake cups themselves could also be described as partially wrapping each other, and the layered cups do seem to signal a greater degree of ritual than the single sake box referred to above. The lacquer, which could be seen as a form of wrapping in itself, is not a vital part of the equipment, but it is also used sometimes for the barrels out of which sake is served and may be seen as adding a ritual layer. Boxes are themselves sometimes arranged to fit inside one another and O'Young Lee (1984) has picked out these 'boxes within boxes' as a peculiarly Japanese device which gives 'concrete expression' to a 'principle of inclusion' which he also finds characteristically Japanese. In fact they are not unlike the boxes described in English as Chinese boxes, but Lee has perhaps picked up the value-creating properties of layers in a Japanese view.

CLOTH

Cloth is of course another substance used for wrapping, as was seen above in the section on *furoshiki*. It is now used in many other ways, some of which may be quite mundane, many protective, and some symbolic. Most people in Japan, especially children, own a large number of bags, for example, and these are perhaps used as much to help keep possessions in order as for any other reason. The practice is learnt early, for schools insist that children bring bags for almost everything they own—indoor shoes, football boots, plimsolls, sports clothes, cleaning equipment, serving-hat and -gown, and pens and pencils—to say nothing of the compulsory satchel to carry books to and fro.

Cloth is also used as an inside layer of protective wrapping for pots, scrolls, and other precious objects. This time it may well be silk which is chosen, the quality of the cloth reflecting the quality of the goods to be wrapped. Many personal treasures may be kept permanently wrapped in special cloths, and a peep inside a household linen cupboard would demonstrate the wide variety of parcels of this sort which are stored. Clothes are often themselves packed up in this way, especially when they are out of season, or during the spring rains, which have a habit of seeping into everything that is not well protected.

Cloth is again often highly valued in its own right. Japanese silk is well known throughout the world, and the most spectacular examples of highly valued cloth would probably be found in some of the elaborate kimonos which have now become *objets d'art*. The use of layers of kimonos will be considered in Chapter 4, but brocade material of other sorts is a particularly effective way of adding a layer of value to something, without any other obvious functional purpose. The brocade 'wrappings' at each end of the musical instrument known as a *koto* would seem to be an example of this. Some of the smaller protective *omamori* are also often contained in a little brocade bag, although the inner part will probably be wrapped in paper (Fig. 2.6).

An interesting cloth parallel with paper and wood, forming aesthetically pleasing wrappings of interior space, is to be found in the *noren*. This is a cloth partition, usually hung up at the entrance to a shop, bar, or restaurant, though sometimes found in domestic entrances too. It has some functional aspects, like keeping out dust and flies during hot weather, but it is patently not only for that purpose since many of them are very short. The *noren* at the entrance to shops and restaurants usually boldly proclaims the name of the

FIG. 2.13. A three-tier set of lacquer sake cups is used for ritual occasions too. See also Fig. 4.15.

FIG. 2.14. This cloth gate-wrapping displays the household crest.

establishment, and this value as a sign would seem to be very important. According to a feature in a Japanese magazine, this symbolic value is paramount to the keepers of such establishments, particularly in communities which like to preserve a traditional Japanese atmosphere (Koizumi, 1985: 5–25).

The hanging of cloth curtains has been a feature of Japanese interior design throughout many periods of history. It was another way in which people could demonstrate their wealth and aesthetic appreciation, as well as cutting out cold draughts or encouraging cool breezes. The word *noren* is said to come from a fifteenth/sixteenth-century Zen Buddhist practice of hanging curtains inside a doorway in the winter. The characters imply that it has a warming purpose, although there seems to have been a cool version for summer as well (Koizumi, 1985: 14). Another variety of cloth wrapping, hung up displaying the household crest over the gates of samurai establishments (Fig. 2.14), must have had a purpose more akin to the modern *noren*, however.

LANGUAGE

Lee's argument above about the Japanese 'principle of inclusion' is based on what he regards as a fundamental difference in the Japanese and Korean languages, despite many similarities between the two. This difference is evident when Japanese people try to speak Korean, and it consequently appears in Korean jokes about Japanese. The essence of the argument is that the use of the possessive form, *no*, is necessary in Japanese when a similar sentence in Korean would not require this part of speech. The use of the particle makes possible a kind of poetry which is difficult to translate, but which Lee argues reduces large spatial expanses of the landscape down to a size with which human beings can cope. As an example, Lee quotes a famous poem by Ishikawa Takuboku which he renders as follows:

Tokai no	On the white sand beach
kojima no iso no	Of a tiny island
shirasuna ni	In the Eastern Sea
ware nakinurete	Bathed in tears
kani to tawamuru	I toy with a crab
	(1984: 25)

Lee argues that the use of the possessive *no* reduces 'the vast, boundless Eastern Sea' to a 'small island', then, further, through 'beach' and 'white sand' 'down to a tiny crab', and then, 'since the poet is weeping, we have in essence the great Eastern Sea in a single teardrop' (ibid. 27). The order of the original Japanese version of the poem, which is not rendered in its English translation, moves from the large through to the small in a way which can plausibly be compared not only with 'boxes within boxes' but can also perhaps be seen as a device which 'wraps' the crab (and possibly also the tear) in layers of the environment in which it is found. The otherwise sad and rather insignificant human being is thus able to take a place in the beauty of the surrounding scenery and adopt an aesthetically pleasing posture which the reader may appreciate.

This poem demonstrates in a medium which would be universally regarded as culturally refined, an aspect of Japanese expression which I would like to argue exists much more widely in language use. This is essentially a way in which language may be compared with material forms of wrapping as we have already described them. There is no doubt that forms of poetry are attributed a value quite aside from the functional, usually described as aesthetic, but an aesthetic element in everyday language is not always as easy to discern, if it exists at all. In Japanese, I would like to argue, there are certain other forms of language, used in everyday life as well as on special occasions, which have an important aesthetic element, just as do Japanese forms of wrapping. These will be considered in detail in the following chapter. Moreover, further qualities of the material forms of wrapping are paralleled, not only in the use of language, but also in other social arenas which we will turn to consider in later chapters.

3

Japanese Language as Wrapping

Remaining for the time being in the Japanese arena, this chapter will attempt to demonstrate the way in which Japanese language may be considered as a type of wrapping. Various examples will be given of usages in material forms of wrapping which are paralleled in language, and it will also be shown how the two are sometimes different aspects of the same phenomenon. It will be argued that this approach to language in Japan may ultimately lead to a deeper understanding of communication in that country since it helps to remove some of the culturally relative assumptions which may otherwise be applied. It should also open the way for the wider consideration of linguistic forms of wrapping.

Honorific Language as Linguistic Wrapping

Japanese lends itself particularly well to this approach because it has clear differences in the use of language for different people, different places, and different occasions. Some of this variety is achieved through the use of a category of forms known as honorifics, or respect language. The Japanese term for these forms is *keigo*, a collective term which includes several sub-categories. The first, *sonkeigo*, a more literal translation of 'respect language', ostensibly raises the relative level of an addressee or referent, the second, *kenjōgo*, or 'humble language', lowers the level of the speaker, and the third, *teineigo*, which is usually translated as 'polite language', in its most straight-forward form raises the general level of the speech altogether. A new category, identified by some linguists, is *bikago*, literally 'beautification language' (Ōishi, 1975; F. Inoue, 1989), which has an effect something akin to adornment of the language used.

These forms of speech are in theory available to all Japanese speakers, and some tuition in their use is given at school. Parents teach their children to speak appropriately to senior members of the family and to teachers, doctors, and other professionals they may encounter together. In the workplace, companies have training courses for their new recruits including instruction in the language they like used with customers. In the kindergarten where I worked in Chiba prefecture, even the teachers were instructed regularly by the head teacher about appropriate forms for the parents of their charges (Hendry,

1986). Nevertheless, manuals about proper usage are available in such abundance in bookshops that it would seem that people still feel they need further guidance. In fact, this language seems to have changed quite considerably in the lifetimes of older people in Japan, who regularly lament the lack of correct usage in the modern world.

In practice, there is a great deal of variety in the quality of this language and in the skill people demonstrate in using it. *Keigo* itself has been described as a 'precious beauty', the 'essence'[1] or 'cream' of the Japanese language (Kusakabe, 1983: 3–5). Many informants, whether they feel able or not to use *keigo* successfully, talk of the beauty of such language when it is used well. Like an artistic form, its beauty can be appreciated even by those who cannot use it. It is said to give Japanese 'a lyrical quality, expressive of feelings', in contrast to the strength of logic in European languages, and without it, it is felt that Japanese would lose its 'charm' (Bunkachō, 1974: 9–10). It is evident, then, that this form of Japanese is valued, somewhat in the same way as the materials used in wrapping are. Let us turn then to examine some aspects of the way *keigo* works.

EXAMPLES OF *KEIGO*

First of all, as in other languages, there are various terms of address depending on a person's occupation and status, and these include some considerable variety in words which play the part of personal pronouns. The second person (or the addressee) is more often addressed by a name and/or title than by a personal pronoun in Japanese, and in fact the second-person pronoun offered in many Japanese textbooks is particularly tricky because it is perhaps most often heard as a rather affectionate form used by a wife to her husband![2] There are other terms which are usually used only by men, and even then, only in informal circumstances. The wrong choice of term could easily sound offensive, ill-educated, or overly affectionate, so this is evidently an area where some care is required.

For self-reference, too, there are different words used by men and by women, and for both men and women, again, these vary depending on the company in which they find themselves. The terms used by women are actually used by men in more formal situations, but the forms used mainly by men are usually only used by women or girls when they wish to make a special point of expressing their inherent equality. There is less variety in third-person pronouns, which are in fact not used that often, one linguist arguing that they give a foreign tone to the sentence, except when they are used rather provocatively to refer to a real or potential lover (Hinds, 1975). In fact, the rest of the syntax of a Japanese sentence often makes the use of personal pronouns redundant.

There are, for example, alternative words for common verbs such as come, go, do, eat, look at, and so forth which add a note of respect for the addressee or a third party mentioned, or have the effect of humbling the speaker. Further distinctions are made based on the choice from a range of verb endings, so that it is usually evident from the verb chosen and its ending whose action it represents. Some examples will illustrate the possibilities. Let us take a simple sentence, such as

where are you going?

For a friend or close acquaintance of an approximately equivalent status, this could in Japanese simply be

doko iku no

which consists of a word for 'where' (*doko*), an infinitive form of the verb 'to go' (*iku*), and an informal interrogative particle. This would sound impolite to a stranger, however, and a more neutral form of expressing the same question would be:

doko e ikimasu ka

This version introduces the correct grammatical particle after the word for 'where', it makes the ending of the verb 'to go' more polite, and it also uses a more orthodox interrogative particle. Although this is a 'safe' form and one likely to be learned by foreigners studying Japanese, it expresses little in the way of social relations, and it is therefore probably not a particularly common form. A more respectful version, which enters the realm of *keigo*, is:

dochira e irasshaimasu ka

which incorporates a more polite version of 'where' and a more respectful form of 'to go' (*irassaru*).

Another respectful form of the verb 'to go' (and 'to come') is *oide ni naru*, which introduces a form of modification to the original verb which may be applied to many other verbs as well, in each case indicating respect for the person to whom the verb applies. There is also a humble (*kenjōgo*) version of the verb 'to go', which is *mairu*, and by using this verb in reference to one's own actions, one is again expressing respect for the addressee, or possibly for a third party or a destination which may have sacred qualities. There is another special construction which may produce a comparable effect in other verbs. This involves an *o* before the verb and the addition of a regular or humble version of the verb 'to do' after it.

It can be seen, then, that there is considerable choice in the use of the verb 'to go', to take only one example. Another commonly used verb is 'to eat', and the various Japanese versions of this illustrate another dimension of the variety involved. There is, again, a standard version, *taberu*, which is the 'safe' one, there is a polite version, *meshiagaru*, which may be used as *sonkeigo*, or just to raise the tone of speech as *teineigo*, particularly by women, and there is a more vulgar version, *meshi suru*, which is generally used only amongst men. Each of these forms can also be used with different endings, which adds another set of subtly different possibilities depending on the intimacy of the speakers.

Another example of the way *teineigo* generally raises the level of speech is the use of *dochira*, mentioned above, for *doko*. It may also be used for 'who', or 'whom', in its plain form *dare*, although this has an elevated form of its own, which is *donata*. *Dochira* can also simply mean 'which?' There are parallel sets of polite and less polite forms of words for 'this one', 'that one', and 'that one by you', which may again (as *sonkeigo*) make redundant the use of personal pronouns, although there is a whole range of variety for these too.

Other commonly used modifiers which may be applied to nouns are *o* or *go* as prefixes on certain words. This again has the effect of raising the general tone of the language, although it may indicate an object belonging to an addressee in order to distinguish it from the general class of such objects. There is considerable disagreement about which words should and should not take *o* and *go* (see Uno, 1985: 20), and in my view this is one of the most basic ways in which people express different social allegiances. A liberal sprinkling of such honorifics may express a good upbringing to some, an excess of femininity to others, and a vulgar attempt to rise above one's background to yet others. Some limited conclusions may be drawn from the words which are honoured in this way, although a number of them would seem to be conventionally shared across the board. The attachment of *go* to the word for 'book' would seem to be an example of the former, however.

This is but a brief introduction to some of the possible variety in the way sentences with similar meaning may be constructed in Japanese, but it should be clear that an expression may come in several possible types of linguistic wrapping depending on the choice of politeness level. In the following sections, there will be a consideration of how choices are made about forms of address, which verb forms and verb endings are to be used, whether words are to be assigned honorific prefixes, and so on.

CHOICE OF LINGUISTIC WRAPPING

In its most direct form, *keigo*, at least in the subcategories of *sonkeigo* and *kenjōgo*, is concerned in all these manifestations with the expression of differences of status. Thus, strictly speaking, one could identify the appropriate language to use, or the appropriate way to 'wrap' one's communication, simply by calculating the relative status between oneself and one's addressee, and between any other parties mentioned or involved in the conversation. There would also be an element of context to consider, as the same people may use different forms in different situations. The office would elicit more formal interaction than the bar, for example, and women are generally more polite to one another when they are dressed in kimono than when they are out on the tennis court (Hendry, 1990*b*: 26). Thus, in theory, if one knows the rules of hierarchy, one should be able to calculate an appropriate form of *keigo* to use.

In practice, however, there are several further subtleties to be taken into consideration, and thus, like wrapping, polite and respectful language is an area which is open to cross-cultural confusion. Japanese visitors to England generally manage to speak and behave in a way which makes them seem extremely 'polite' to the people with whom they come into contact, and they are therefore regarded as pleasant people to deal with, unlike visitors whose languages lack words for 'please' or 'thank you' in approximately similar contexts. However, this politeness may well be covering up something which would be better brought out into the open in some circumstances, and long-term grievances may build up because of subtle differences in the rather similar wrapping forms used. Generally, I would argue, there is a lack of fit at the non-verbal level in the communication of politeness in Japan and Britain, but there are also many comparable features.

One example of the latter is the way polite language, including the respectful and humble forms, is applied, literally to oil the works. This expression covers several types of interaction. For example, it may be used in order to make unpleasant or unwelcome news more palatable, and to soften the blow of revealing such communication. It may also be used when asking favours, or requesting something which could cause the recipient of the request some inconvenience. It is thus similar to a sort of flattery, found in many languages, where a person complies with a request because it makes them feel good to do so. This may, in particular, be a way in which subordinates exercise some power over their superordinates, and the dyad can apply well to female/male interaction too. At a commercial level, this strategy may be used to persuade customers into buying goods, and a classic example is discussed in some detail in a Japanese book on the subject of *keigo* (H. Araki, 1983: 36–50).

The case concerned a Kyoto man who gave up a secure job in the City Hall to set up his own confectionery business. He decided to specialize in traditional Japanese cakes made from the finest ingredients, and he gradually built up such a successful business that he became known as the number one confectioner in the land. When asked to explain his success, he mentioned, of course, the excellence of his ingredients, but he went further and explained his philosophy of caring for his customers. He emphasized the importance of using polite language with them, and also of meaning sincerely the words which he and his staff used to address them. The words express care, and they must show care, he argued, claiming that the real reason for his success lay in the true feeling he put into the etiquette he used.

Keigo may also be used to introduce an element of social distance into discourse, however, and the converse is also true, whereby people drop the polite forms they have been using in order to express some intimacy. This distinction is directly comparable with material wrapping, where gifts to intimate friends and relatives may be presented completely without covering as an expression of social proximity. The formality of polite speech is usual in conversation between strangers, particularly if the meeting holds the expectation of longer term interaction, but these formalities may be gradually dropped as the participants become better acquainted. Particularly in the case of women, whose status may be approximately equal, formal language will initiate a relationship, but continuing intercourse will almost certainly lead to greater intimacy and informality.

The use of *keigo* then becomes part of a subtle means of signalling between friends and acquaintances, as each side may put on the brakes if they feel intimacy is escalating or one side is becoming too informal, simply by stepping up their own use of *keigo*. This has the effect of sounding cold and unfriendly in a situation where more informal communication has been the norm, and the recipient of such a put-down may well examine their own interaction for signs of too much intimacy. A hospital matron who became involved in my research on this subject noted that nurses should avoid being too polite with patients because of the distance which this may set up between them. A result of this approach, however, is that some patients complain of being treated like children since they receive less than the usual respect they are accorded in the outside world (Hendry, 1990*c*: 114).

The same mechanism can be put to good effect purposely to discourage advances which are unwanted. For example, housewives will speak in a cold, but polite, tone to door-to-door salespeople in order to put them off, and to indicate that they have no interest in their products. Similarly, people can politely discourage all kinds of approaches which they find inconvenient, embarrassing or just unpleasant without any need for outright rudeness, except that which is implied by too great a use of *keigo*.[3] In this respect, too, English has polite language available for such effects, although it is perhaps not used as much as it used to be. Some of Jane Austen's characters were mistresses of the art, however! (cf. Segal and Handler, 1989.)

In Japan, too, there are regional differences in the expected use of *keigo* (e.g. Iitoyo, 1966; F. Inoue, 1989). During the research period when I was investigating this subject I was living in a provincial town some two hours by train from Tokyo. Amongst my close associates were women who had been brought up in the area as well as women who had moved into the area from more distant locations, including Tokyo and Kyoto. Some of them talked to me of the regional variations they had encountered, and one or two made conscious efforts to modify the language they had used in other places where they had lived. In some parts of Tokyo and Kyoto, for example, a much greater degree of *keigo* is expected in everyday conversation than is the case in this provincial town.

Amongst these women, however, were two who confessed that they found it extremely difficult to adjust to the area. One even complained that she felt constantly as if she was living in a foreign country. I noticed that in both these cases the women were continuing to use the very polite language they had been brought up to use in Tokyo, and I feel sure that this at least in part explained their problem. To their new neighbours and associates they simply sounded unfriendly, possibly even 'stuck up'. A third woman, one of those who told me that she had felt the need to adjust her language, was highly regarded by her neighbours, one of whom actually commented to me that she was surprised by how well this Tokyo doctor's wife seemed able to speak the same language as they did.

RITUAL WRAPPING

In the most rural regions of Japan, respect language is little used in everyday life, for people will be dealing with close acquaintances most of the time, and their mode of discourse may well not include much *keigo* at all. In a similar way, everyday exchanges of produce, or surpluses which accumulate in the home, will not be elaborately enclosed. Formal wrapping will be reserved for special occasions, when it will add a ritual air to the occasion, and this applies both to the presentation of gifts and the presentation of language. Thus, for example, a wedding, or the birth of a baby, will be the occasion for the exchange of properly wrapped gifts, and for much formal communication of stereotypical phrases of greeting and well wishing.

In Kyushu where I carried out fieldwork on weddings in the 1970s, for example, there were several presentations of gifts which preceded, coincided with, and followed the nuptials, and each of them had appropriate forms of speech to express the sentiments of the occasion.[4] First, there was a presentation of tea, known either as 'decide tea' or 'nail tea' to express and clinch the decision which the couple and their families have made to be joined in wedlock. The gifts were tea, sake, and a sea bream, each having a symbolic meaning for the occasion, but the tea, in particular, came in a box decorated with an elaboration quite out of proportion to its material value. This is partly so that various other appropriate aspects of the exchange and its future expectations may be symbolized, but mainly simply in order to convert this mundane commodity into a material representation of the occasion being enacted. (A highly abbreviated version is shown in Fig. 3.1 where packaging for an engagement may includes symbols of good fortune and longevity).

Further gifts, infinitely more expensive and elaborate, were handed over from each family to the other a few days before the wedding itself. From the family which will receive a bride (or occasionally a groom), these gifts are in this area referred to as 'main tea', although the tea represents only a small part of the goods involved, and it is even said to be poor and therefore rather cheap tea, for symbolic reasons[5] (Fig. 3.2). There is often a substantial sum of

FIG. 3.1. Packaging for an engagement ring, displayed in a Tokyo department store. See also Pl. VIII.

FIG. 3.2. Packaging for betrothal tea (Yame, Kyushu).

money, or a set of fine garments, but again, the wrapping and the abundance of symbolic gifts, is just as important. The gifts travelling the other way are somewhat more functional, in that they generally include large items of furniture or equipment for the future lives of the couple, but they are also decorated and 'wrapped' in appropriate ways, particularly a mirror for the bride which is said to represent her soul.

Further lesser exchanges continue throughout the nuptial season, and in each case an element of the material wrapping indicates clearly the nature of the occasion and may well symbolize more specific aspects of the expectations involved. In precisely the same way, the language used as all these gifts are handed over is equally elaborate and formulaic (cf. Bachnik, 1986: 65–6). It usually consists of a series of expressions, which bear little relation to the actual experience or feelings of any of the individuals involved, but which are deemed appropriate for the occasion, and which express the long-term relationships being set up and some of the expectations involved. On the occasions when the biggest gifts are presented, the bride and groom may not even be required to attend. The whole affair is set up and transacted by the elders in their two houses, who have probably paid for much of the procedure, and who will also be linked in future by the match to be enacted. They are also more likely to be experienced in the use of the language appropriate for these occasions!

In the same way, the gifts and language used at any number of ritual occasions will reflect the formal aspects of the proceedings rather than the individual feelings of the participants, and on some occasions the movement of a gift in the appropriate wrapping will be a sufficient expression of the words which could be spoken but which may not be necessary. The gift itself will carry, symbolically, the meaning to be expressed, made quite clear and precise in the wrapping used. After a wedding, for example, the guests will take home gifts prepared in advance. These gifts invariably include foodstuffs, possibly cakes, but almost always decorated in a way which symbolizes the occasion. The guests are thus able further to share the celebration with members of their own households who may not have been able to attend.

Other household ceremonies, perhaps for births, deaths, stages of the development of children, or the reaching of certain ages of the old people, are also shared with neighbours and other associates through the moving of material objects, very often food. The food itself is usually marked with a character for celebration, or condolence in the case of a death or a memorial for the dead, and the wrapping will also indicate clearly what the occasion is. These gifts are usually delivered, with appropriately 'wrapped' phrases of presentation, but the way they carry the message intrinsically allows the appropriate information to be passed on to all the parties who may receive such objects. Phrases of thanks will again include appropriately wrapped phrases.

In the light of these ritual parallels in the use of *keigo* and the wrapping of gifts, an interesting contrast is found in the offerings made to the spiritual world. In Shinto rites the presentation is usually a selection of goods vital to sustain human life. Typically they include one or more of rice, vegetables, fruit, water, and sake, although specific rites will require particular goods. For example, the ground-breaking ceremony for the building of a new private house should apparently include the offering of a pair of fish, to represent the fertility of the household in the future. A harvest thanksgiving may require a selection of grains and seasonal produce. Offerings made at the household Buddhist altar will usually include something the most recently deceased member particularly liked.

One common feature, however, is that these goods should be opened up and presented, unwrapped, to the deities or ancestors concerned. They may be placed on a special offering table, or decanted into an offering bottle or barrel in the case of sake, but they should properly be open and visible (Fig. 3.3), according to religious specialists I consulted on the subject. This practice is not always observed, perhaps especially when a bottle of sake is the item in question, but this is sometimes due to the ignorance of the people making the offering, according to specialists. In the home, the first of the day's rice will be

FIG. 3.3. Offerings to deities are usually unwrapped.

served, open, to the ancestors, and fruit and other goods may be laid on the offering table.[6]

It was particularly interesting to learn from some Shinto priests at a high-ranking shrine, then, that when ordinary people speak to the deities, they are not expected to use *keigo* or polite language at all, rather they communicate their prayers, or their wishes, through what the priests described as *kotodama no shinkō* or 'communication from the heart', which may not require any formal articulation at all (see Hendry, 1990*b*: 29). Like the offerings they make, they present themselves entirely unwrapped, free from any of the social packaging with which they need to confront other human beings.

LANGUAGE AS A FORM OF SELF-DEFENCE

Formal language is generally used when people are 'on their best behaviour' for some reason. In Japanese these are called *aratamatta toki*, and they include a wide variety of events. Basically, they are occasions when people adopt a special posture, or *kamaeru* in Japanese, an expression which has a connotation of self-defence. It is used, for example, for the adoption of postures found in martial arts, and for 'getting ready' for a battle or a match in a sport such as tennis or baseball. In this way, *keigo* may be described as a kind of armour, to wrap up and protect the nerves which may be lurking underneath, and, as described above, to put some deflecting distance between oneself and a threatening world which one may encounter 'out there'.

The notion of 'battle' in connection with *keigo* is not purely metaphorical, in fact, because there are veritable battles continuing every day between people who use language in order to gain an edge over those around them in constant struggles for ascendance in the informal hierarchies of status and prestige. Particularly among women, language is used, as elsewhere, as a means of assessment of factors such as upbringing and education, and those who wish to impress their fellows in this respect may pay special attention to their use of *keigo*. It is precisely in this area that people judge one another's use of *teineigo* and *bikago* for being more or less in accordance with what one accepts oneself as right and proper, although of course such judgements may be perceived as being made against more universal standards of correctness, even if these standards are objectively difficult to identify.

Many of the Japanese women I worked with on this topic were quite happy to pass judgements on other people around them, and those who regarded themselves as skilful in the use of *keigo* explained that they could identify whether or not a person had been brought up since childhood to use the 'proper' language (as they perceived it), or whether they had picked it up or been taught to use it later in life. Actually, some of these women told me that they dropped the most polite forms when they were in the company of people who were less skilled so that, in effect, an ability to use the most polite forms well not only acts as a kind of passport into the company of those who share this skill but also excludes others by denying them access to its use.

Rather paradoxically, then, those who are less confident are the ones who are always trying to keep up the polite patter (cf. Tanaka, 1966: 159). Sometimes women from the country who have moved to Tokyo fall into this class, and one of my informants described her own mother-in-law, who was from Tōhoku, as 'standing on her tiptoes' to get her *keigo* right. A satirical onomatopoeic name for the pseudo-upper-class, which developed in the early decades of the century partly by using the speech of samurai families as their model (ibid. 156), is the *zaamasu* brigade, a word with connotations of the effort involved in all the *gozaimasu* polite expressions they are said to use.

The level of *keigo* ability acts as another kind of wrapping, then, separating off the class of skilful users from those who are less accomplished. In this respect, it resembles the use of a local dialect, which serves the same function of separating off local users of a language from outsiders who are unused to its idiosyncrasies. At the higher echelons of society, language becomes very esoteric, so that most ordinary Japanese would find it very difficult to converse at all with members of the imperial family, and this is a way in which they are protected from the approaches of the common people. In later chapters we look at other forms of wrapping which play this role for people in positions of great status, indeed, we examine some of the wrapping as communicating that status.

In a more positive manner, *keigo* is also a means by which people can demonstrate their taste and preference in the use of language. With skill, they can wrap themselves according to the image they want to present to the world and the persona they want to express. This linguistic wrapping is not only an appropriate means of concealing less worthy thoughts, which may prove a surprise for those who break through the polite layers, but a means to develop the kind of Japanese refinement which is characterized both in this mode of communication and in the attention paid to the wrapping of gifts. This is, of course, the aesthetic quality of the language which we began to discuss at the end of the previous chapter as another way in which some of its uses may be compared with the beauty of material wrapping. However, just as Oka (1988) lamented a recent reluctance to devote time and care to the creation of beautiful packages, there are those who lament the passing of the beautiful language they used to know in the rush of efficiency which has overtaken the modern world.[7]

MATERIAL AND LINGUISTIC WRAPPING AS EXPRESSIONS OF CARE

In Chapter 1, it was mentioned that layers of wrapping are a way of expressing care for the object inside, and therefore care for the recipient of the object.

This was written in a book on the subject of *keigo* by a Japanese linguist, Yoshikata Uno (1985), who also sees the use of *keigo* as a way of expressing care for the person one addresses. In every case the actions are examples of politeness, where the Japanese word for this concept, *teinei*, can also be translated as 'care'. Another example is the use, already mentioned, of wrapping a single-page letter in a plain sheet. Uno also refers to people who talk of the 'magical' properties of *keigo* to smooth over unpleasant and difficult events (ibid. 47), but he warns that too much *keigo* can have the opposite effect in creating a feeling of distance between the two people involved in an exchange (ibid. 42).

Another Japanese word which is used in the context of polite language and behaviour combines these two notions of care and distance. It is *enryō*, which may be divided into *en*, meaning 'distance', and *ryō*, which may be translated as 'consideration'. This word is generally used in efforts to put people at their ease. *Enryō shinaide*, or 'don't do *enryō*', means something like 'you don't need to be on your best behaviour', or 'please relax'. It is also used when a person feels another is holding back for the sake of politeness, and not revealing their true wants or needs, perhaps in consideration to their host. In this way, it implies that the polite behaviour, in particular language, is covering up some physical desire which could be expressed in more informal circumstances. Thus it is a form of wrapping which expresses care and consideration, but also an element of distance.

In general, it is felt to be imposing on people to reveal too much of one's own feelings, and it is also something of an art, particularly among women, to be able to anticipate or divine what lies behind a mask of polite expression. There are certain groups of people, such as relatives and close acquaintances, with whom one would expect to share confidences, and these would also be the people whose gifts would normally require little wrapping. There are also times and places for self-revelation, just as there are times and places for formal speeches and beautifully wrapped gifts. In this sense, by expressing consideration, politeness, and some appropriate sentiments, linguistic and material wrapping may again be doing much the same thing.

Language with Content Concealed in the Wrapping

In the first part of the last chapter, it was pointed out that the towels and sheets used in token gift exchange between houses could perhaps be seen as the gift and wrapping all rolled into one. Indeed, the gift is less significant than its movement, so that the towel might even just represent wrapping. In a similar way, there are forms of speech which have so little content that they almost become significant only in the wrapping they represent. Sometimes such speech does have ostensible content, in that the words used may have literal meaning which could conceivably be taken at face value, but usually the context makes clear that they are merely appropriate words for the occasion—perhaps expressing care and consideration again.

Examples are when speeches are made to gatherings of people, or when words of formal welcome are spoken to visitors on a ritual occasion. One favourite expression is to thank people for giving up their time when they are so busy, others refer to the cold/hot (according to the season) weather the visitors have had to endure in order to be present. The impression given is that those being addressed have in some way suffered and sacrificed for the occasion, even if it is being held principally for their benefit and they have almost certainly chosen to be there. It is simply a part of general politeness to express care and consideration for those with whom one interacts, and it is even possible to thank someone for their politeness rather than for the material gift they have actually handed over.

Another example of this type of language may actually be found in expressions conventionally used in the handing over of gifts. The phrase used by the giver of a gift generally plays down the value of the object, even (and perhaps especially) if it is an object of some considerable worth. *Tsumaranai mono desu ga* runs the phrase, 'it's not worth having, but . . .'. A similar expression describes the apparent paucity of fare when making an invitation to someone to share a meal. This time the phrase, *nanimo nai desu kedo*, is literally pronouncing that 'there is nothing at all available, but . . .'. These phrases are evidently not supposed to be taken at face value, and the content of the wrapping is perhaps again, as in the case of the towels, expressing a reluctance to impose an obligation.

In fact, language which is almost totally incomprehensible to the ordinary participants is used in much religious ceremony in Japan. Shinto prayers are chanted in a language of worship developed many centuries ago, and the Buddhist sutras, though they are often learned by heart, are not in Japanese at all. Some of these convey some meaning in their Chinese written form, others are phoneticized Sanskrit, but with pronunciation shifts which makes them incomprehensible to all but the most erudite Buddhist scholars. The value of this kind of speech is entirely symbolic to the lay participants and it is the form of the chanting that signifies its importance rather than the content of the words themselves.[8] This type of utterance would fall at one extreme of a continuum which could be drawn up with form at one end and content at the other, a continuum which could perhaps also be represented as a layering of wrapping with increasing value placed in the type and amount of wrapping rather than any intrinsic meaning it may contain.[9] Instead, the meaning is to be found again in the wrapping and its appropriateness to the occasion.

This continuum is a little reminiscent of a scale of 'formalization' of language drawn up by Maurice Bloch in a discussion of language used in ritual, during which he assesses the extent to which individual creativity is possible in the uses considered. He comes to the conclusion that in the case of the ritual of the Merina people of Madagascar, singing comes at the most fixed (or wrapped, to use my terms) end of the scale (1974: 70). Although I don't necessarily endorse Bloch's argument in its entirety here, he makes the interesting suggestion that words used in this way in a ritual context are acting more like things, like material symbols, than ordinary language (ibid. 75).[10]

At this point, we might press a little further with a few cross-cultural comparisons, for it is by no means unique to Japan to carry out ceremony in a language beyond the understanding of the participants. The use of Latin in European churches no doubt had an effect for the lay participants rather similar to that described above for Shinto and Buddhist chants, and the Latin degree ceremony persists in Oxford despite the ever-decreasing number of students who have more than a smattering of knowledge of it. The Latin does of course carry meaning closely related to the event which is taking place, at least at a formal level, and there are those present who understand the purpose of its content.

For other participants, however, the meaning of the language is to be found largely in the formality and ritual it imparts to the occasion, and those who lament the passing of the Latin Mass, or even the earlier English translations of the Bible, are almost certainly lamenting the loss of the effectiveness of the wrapping in which the words are presented. The throwing out of relatively incomprehensible aspects of religious ritual may perhaps here be compared with pleas to do away with the apparently unnecessary layers of material wrapping in the interest of conservation of resources. In each case, there is a perfectly reasonable functional case to be made, but they both neglect the symbolic power of the wrapping involved.

In the Japanese case, I would argue, few would doubt the power associated with wrapping. We have seen the important roles played by the use of *keigo*, but this may well have been influenced by religious ideas. An interesting example, which crosses the boundary into everyday life, is to be found in the power attributed to words by followers of a relatively new Buddhist sect known as Mahikari. It is reported here that 'clearly articulate chanting, loud energetic greetings, and addressing others (and things) in a warm, friendly manner has a beneficial effect' (McVeigh, 1991: 133). These forms of speech are referred to as *kotodama* ('a spiritual power of words') and doses of this can apparently help alleviate family discord, educational problems, or work-related stress, as well as encouraging the growth of plants (ibid.).

In a more strictly religious context, though still in Japan, Carmen Blacker analyses the power of words—to 'cure sickness, overcome demons, vanquish enemies, cause rain to fall and children to be conceived' (1975: 93). She emphasizes again the way the power of words may lie more in their form than in their literal meaning, particularly in the case of the mantras and *dhārani* of esoteric Buddhism, since very few people understand their meaning.[11] She makes the point that the sounds themselves can have two kinds of power, first to bring about the desired effect in the world and secondly, under certain conditions, to create power in those who recite them.

An ability to use esoteric language confirms for a person a special role in society, whether it be as priest or proctor, just as it does for Shinto and Buddhist priests. Elsewhere, other forms of ceremonial language mark off ritual, political, and religious leaders from the rest of the populace, as well as marking an occasion as one of ritual importance. Another good example is to be found in the 'ceremonial dialogue' used among some South American Indian peoples. Here, again, certain elders only are able to converse in the most special language required, in its most powerful form typically used to arrange tricky negotiations such as outrageous trade offers, peace treaties, and marriage alliances.

In his analysis of ceremonial dialogue amongst the Trio Indians of Surinam, Peter Rivière (1971) suggests that the practice is used as a form of mediation in situations likely to give rise to conflict. According to his report, there are different strengths of ceremonial dialogue depending on the occasion, and the stronger the type, the fewer the people able to use it. This form of dialogue is used only with outsiders to the immediate residential group and the strength of the dialogue used is directly proportional to the social and physical distance of the partners engaged in the dialogue. In this way, then, the practice also resembles the use of *keigo* where levels of politeness increase with social distance and formality, and may well increase in proportion to the difficulty of a problem to be discussed.

In the case of *keigo*, too, as noted above, its aesthetic qualities ascribe to those who can use it well a certain eloquence which serves in many cases also to mark them as members of a rather élite part of society, something akin to, though not necessarily the same as, the nobility in Japan. The nobility do use and expect extremely polite language, indeed the funeral of the Emperor Showa taxed even the most accomplished journalists to find suitable language to refer to him, and this marks members of the nobility off from other members of Japanese society. At a level slightly lower than this, there are people who are socially very accomplished, who regard their use of *keigo* as a vital skill to pass on to their children, and these form another part of Japanese 'high society'.

All these examples illustrate the power which an ability to use certain forms of language may hold, particularly where this ability is restricted to closed groups of people. This power is derived partly from the status ascribed to the linguistic wrapping involved, and in some cases, such as that of the Japanese emperor, involves little more than status. Real power I would suggest lies rather in a combination of abilities: first, the access to different forms of language, and, secondly, an ability to manipulate the use of those linguistic forms. This is a subject to which we will return in more detail in Chapter 8.

Cross-Cultural Confusion

Linguistic skills are undoubtedly important in all societies, and in complex ones they take on roles of social distinction (cf. Grillo, 1989). However, linguistic subtleties such as these are extremely difficult for non-native speakers to understand and even more difficult for them to use, particularly if they are automatically classified for some other reason, such as skin colour, as a foreigner. Moreover, when assumptions about linguistic communication in one society are transferred to other languages in other societies, they may be misinterpreted by the native speakers of that language, and this is what I suggest sometimes happens when Japanese native speakers transfer their notions of politeness to English, or, indeed, any other language.

Conversely, non-native Japanese speakers could find themselves in a minefield of difficulties in Japanese if they were really expected to communicate on the same level as native speakers, which for the most part they are not. Indeed, Japanese politeness is such that most foreigners studying Japanese are led to believe that they are doing an awful lot better at the language than they usually are, simply because they are doing anything at all. Moreover, the better they get at the language, the less the Japanese themselves tend to approve, and one distinguished Japanese linguist of my acquaintance claims that he finds it a kind of psychological torture to listen to foreigners speaking Japanese, even 'perfect Japanese', because he can't understand their 'real intent'.

I have discussed some of these difficulties in more detail elsewhere (Hendry, 1990*a*), particularly as this is something of an occupational hazard for ethnographers, but it is pertinent to note here that this particular linguist holds rather idiosyncratic views on the subject of foreigners learning Japanese. Nevertheless, it would seem to be the layers of linguistic wrapping which he feels are somehow being misused, or perhaps he just cannot believe that they could be properly used by foreigners. His ideas, like those of many other Japanese, betray a conviction that his language is somehow unique, which, I suppose, in the sense that any language has unique qualities, it may be.

The problems exist at some level, however, for everybody communicating with people speaking a language other than their own, although both parties may be blissfully unaware of them. They involve not only politeness, but attitudes to other aspects of indirection such as deceit, joking, allusion, and so on. In the case of language, these are issues which have been discussed by sociolinguists, and language courses are beginning to devote at least a small part of their time to such cultural issues, but this aspect of wrapping is not limited to speech forms. Instead, in different degrees in different languages, non-verbal cues play a more or less important part which may be realized only to a limited extent, especially by people whose own language is verbally explicit. In the next two chapters of this book, some of these non-verbal cues will be considered in detail, both in the Japanese case, and more generally.

4

Wrapping of the Body

While the spoken word is undoubtedly a very powerful means of communicating all kinds of non-verbal information, it is at the same time limited as a means of such communication to those who are well versed in the use of a particular language. The presentation of gifts, on the other hand, has for long been a means of making contact between different linguistic groups, and eventually, long-term communication between such groups may be extended to trade and political interaction. At this stage, groups enter into complicated relations of competition with one another. People recognize the need to learn each other's languages, but they may be less aware of the potency of other messages. This and the following chapter will consider two other forms of wrapping which are variously deployed in communication both within and between cultures.

First, this chapter will be concerned with the wrapping of the human body. This is one of the most obvious ways of communicating information about ourselves, or reading information about others, although we may not do these things entirely consciously. As fashion designers have long been aware, what we wear and how we wear it can greatly influence the impression we make on other people. Perhaps without really thinking about it, we already tend to make assumptions about other people when we have done little more than catch sight of their outward appearance. Within our own society, these assumptions may be fairly accurate, even quite useful, but in an intercultural context, it is possible to be guilty of quite disastrous errors of judgement in this way.

This chapter will look, first, at some general ways in which people size one another up, based on their dress and other aspects of their bodily wrapping, drawing largely on the expected experience of the Western reader. To give some examples of quite a different code, it will then turn to look in some detail at examples of Japanese bodily wrapping, an area which offers tremendous variety of symbolic expression, and which opens the way to the final section in which some problems of intercultural communication at this level will be raised.

The Leopard and its Spots

A leopard is a leopard and has little choice about his spots, but human beings have managed to create great variety of wrappings for themselves. These may express in an almost infinite number of ways further information about the being inside, or, alternatively, they may conveniently conceal such information. In the modern world, a man or woman in fairly smart Western clothes, whatever their hair colour or skin type, can travel rather freely in a variety of countries of the Western world, buy almost anything they like with the aid only of a piece of plastic, and attract little further attention. The same would not be true of a man wearing a feather headdress and a penis sheath, nor would it be easy for a woman dressed only in a grass skirt. Despite commonly held ideas about the liberal views towards apparel held in our society, we do have certain limits.

Sitting in a crowded bus, or a tube train, surrounded by strangers, it is quite interesting to play guessing games about the other passengers. What sort of family do they come from? What is their likely occupation? Would we want to get to know them better? The last question is probably rather easier than the first two, although one may be hard pressed to say exactly why. Age may be something to do with it, but you would hardly pick every person of your own age from the crowded train to be a potential friend. You may like or dislike the person's face, their hair colour, their stance; more than likely, however, you would also pay some considerable attention to their clothes, their shoes, their hairstyle, and possibly their make-up. You would probably notice, though perhaps even only subliminally, whether or not they looked clean, tidy, and fresh, or dirty and dishevelled.

A person in a smart suit may look trendy and up to date, one of the jet set, or they may look old-fashioned and dowdy, a remainder from a past period of good fortune. A person in denim jeans, that now almost universal form of wrapping, may look dressed up and 'tarty', with high heels and trousers too tight in the hips, or they may look casual and comfortable. Recently, they may feel trendy by having their jeans split into a number of tears and holes, a condition which their mothers may well rightly have regarded as characteristic of the poor and underprivileged. Fashions change, of course, and so do assumptions about clothes. Depending on the differing views and perspectives of the viewer, the people they view may look appealing or quite off-putting and unpleasant.

There are of course members of modern society who take a lot of trouble to create a particular impression. Using make-up and hairspray is one way to spend time on one's appearance, and the result may be subtle and attractive or stark and tasteless. Wearing expensive clothes is a way to spend money. Jackie Kennedy was a nightmare to those in charge of the White House budget when she was First Lady, but her appearance made a lasting and unforgettable impression on the world. Some modern youths, on the other hand, go out of their way to appear threatening and unpleasant. They wear big black boots and heavy chains, shaving their heads, or arranging for their hair to stand up in styles which resemble the comb of a cock, a Mohican Indian, or a person who has just received a severe shock. Punks know well and enjoy the impact of extraordinary looks on the world around them.

Hats are useful objects for transmitting information about their owners. Although both may be Scottish, a deer-stalker tells quite a different story to a tartan cap with a peak at the front, just as a top hat will convey a view different to a bowler or a beret. Some information about headwear may be evident in different ways to different viewers. A skull-cap to a non-Jew will tell only that the wearer is almost certainly a Jew, probably orthodox. Within a particular Jewish community, however, the skull-cap can convey considerable further information to and about its wearer. According to Suzanne Baizerman (1991), these small items, where possible carefully crocheted in the home, allow women great scope to transmit messages, both to their menfolk for whom they make them, and to the wider world about their menfolk's caretakers and loved ones. As a corollary of this, women are apparently also quite prone to judge one another on the basis of the skill they witness in these rather neat exhibitions of their handiwork.

In other parts of the world, other garments convey similar sorts of information. A man in home-knitted socks usually demonstrates the presence of a caring woman somewhere in the background, as does a cricketer in a home-knitted sweater. In both cases, too, the skill of the craftswoman is on display to the world. In some parts of Scotland and Ireland, home-knitted sweaters also carry the identity of the families of their owners in subtle differences in their colours or patterns, although to an outsider they merely look like local (perhaps ethnic) sweaters, a nice present or souvenir to take back home. Kilts and tartan scarves and skirts have similar contextual meanings. To many, they are just attractive garments of Scottish origin; at a dance or a ceilidh, they allow newcomers to identify husbands and wives, or members of more extended families. In Spain they seem to have become rather popular for girls' school uniform.

Further conclusions may be drawn about the state of clothes. During a recent visit I made to Greece, a woman with a full-time job told me that she still feels obliged to iron all her husband's shirts most carefully. A man with a creased shirt looks neglected, she explained, and it is his wife who will be blamed. Other men may not mind, or even notice, but other women will be sure to see if she misses a shirt, and gossip will be rife! If a man is dishevelled enough, he could be regarded as fair game for another woman to woo, even if he is married, for he is obviously not well cared for. This remark reminds me of my childhood, and my mother's constant pleas for me to wear only clean and tidy clothes. 'People know who you are,' she would entreat, 'you mustn't let the family down.' The idea is summed up in reference to the Sakalava people of Madagascar, 'The properly social person is clothed, and the clothing itself conveys the nature of people's affiliations and their quality' (Feeley-Harnik, 1989: 79).

Japanese Bodily Wrapping

Japanese kimonos, perhaps more than any other garments,[1] are literally 'wrapped' around the body (Fig. 4.1), sometimes in several layers, like the

FIG. 4.1. This Kuniyoshi print (*c*.1842) of a bride being dressed illustrates the way kimonos are literally wrapped around the body; © The Trustees of the National Museums of Scotland 1993. See also Pl. IX.

gifts, and they are secured in place by sashes, with a wide *obi* to complete the human parcel (Fig. 4.2). The *obi* is itself secured with complicated ties, the precise form depending on the occasion, but the final effect intended to offer an aesthetically pleasing overall image to the observer, just as is the *mizuhiki* which secures gifts. Indeed, a book of etiquette considers the two subjects in the same section (Ogasawara, 1985). This similarity with other forms of wrapping has almost certainly influenced the comparisons made in this book, but there are also some interesting correlations to be made between forms. Language, for example, may well alter depending on the clothes being worn. A woman in a kimono is quite likely to raise her level of politeness, for example, whereas one in a skimpy tennis skirt may feel much freer to be

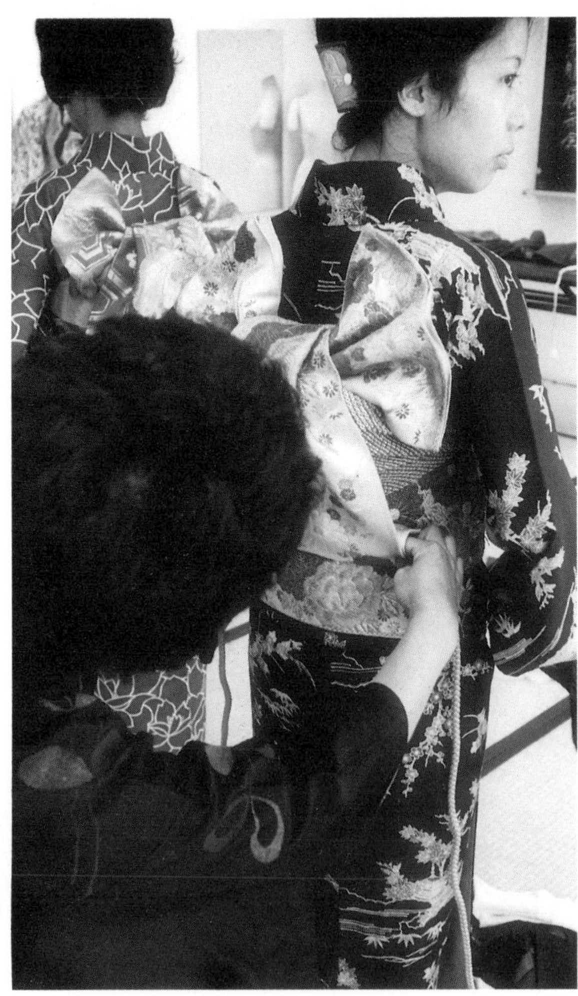

FIG. 4.2. An *obi* must be carefully tied; courtesy Japan National Tourist Office.

relatively informal, even quite coarse and childlike (see Hendry, 1990*b*).

Probably the most extreme Japanese example of the literal 'wrapping' of the body is a set of garments known as the *jūnihitoe*, worn by court ladies of the Heian period (ninth–twelfth centuries). The Japanese word means twelve layers, but in its most elaborate form could describe a garment composed of up to twenty distinct kimonos, each chosen carefully to create together an aesthetically pleasing combination of colour contrasts at the neck and sleeve (see e.g. Dalby, 1988). The wearer of such a garment had very little freedom of movement, and she would sometimes be further obscured by being obliged to remain behind a screen during conversation (whence only her sleeves might be visible), but she would grandly symbolize the wealth and status of the courtly company of which she was part. These garments can still be seen in museums, but imperial weddings are probably the only occasions formal enough for such sumptuous attire actually to be worn (Fig. 4.3).

Most brides in Japan wear rather elaborate garments, as they do elsewhere, and it is customary for them to change two or three times during the wedding reception. The outer kimono usually worn for the actual marriage ceremony

FIG. 4.3. The twelve-layered *jūnihitoe*, worn by the bride of Prince Aya, younger brother of the Crown Prince; courtesy The Imperial Household Agency, Japan. See also Pl. X.

(Fig. 4.4) is another ancient garment of the noble classes, again so heavy and elaborate that movement is severely restricted, and the bride can do little more than stand and be admired! Under this she may have two or three inner layers, one of which will be a pure white kimono, the mark of a rite of passage also used to wrap babies, pilgrims, and corpses. It is said to indicate a *tabula rasa* or 'clean slate', in this case that the bride should bring to her new home. The bride's headdress also carries some traditional symbolism. It is called a 'horn-hider' which demonstrates that she will show no jealousy to her husband (Fig. 4.5). An alternative, older headdress hides the bride's face altogether until she reaches the moment of union, a device which has led to some deception, it is said.

FIG. 4.4. A bridal *uchikake* outer kimono, on display in a Tokyo department store; these garments are usually hired. See also Pl. XI.

FIG. 4.5. A bride and groom at their wedding; courtesy Japan Information and Cultural Centre.

The garments into which a bride changes during the ceremony may have been given to her in the betrothal gifts from her new husband's family. They are said to show her off in different lights. These days brides usually also change at some point into a Western dress, a formal white wedding-dress or, sometimes, a long coloured evening-gown. The bridegroom will then change into a Western suit. His formal Japanese garment will usually have a black coat sporting the crest of his household at various strategic points (Fig. 4.5). The married women at the reception also very often wear formal black kimonos, imprinted with the household crest.

These garments indicate extremes of formality which are also evident in many other areas of Japanese society, albeit involving more subtle levels of symbolic communication. In pre-modern times certain types of kimono were reserved for certain classes of people, and severe punishments awaited any who transgressed the rules. Details such as colour, length, and breadth of sleeves indicated information about the age and social or marital status of the wearer, and many garments sported a set of family crests to make clear even more specific allegiances. Headdress and hairstyle or cut also indicated social divisions, and a moving scene in a dramatic depiction of an inappropriate love affair in the film *Shinju* (Double Suicide) shows the protagonists let down

FIG. 4.6. Kunisada print (1858) showing a courtesan with attendant; © The Trustees of the National Museums of Scotland 1993. See also Pl. XII.

FIG. 4.7. This Kuniyasu kabuki print (1820) illustrates the effectiveness of clothing of a male actor to play a female role; © The Trustees of the National Museums of Scotland 1993.

their hair together to shake off their social ties before they take their lives in a double suicide.

Amongst those classes who had wide resources for acquiring clothes, there has also for long been a tradition and something of an artistic skill in Japan for choosing garments appropriate to an occasion, complete with all its undercurrents. At the most basic level, a kimono should be attuned to the season. First, the weight of the garment is decided rather by the specific date than by the actual climate of the day—a practice which continues in modern Japan—but there will also be more subtle indications of seasonal variation in the colours and decorations on the cloth. Various degrees of auspiciousness, or perhaps condolence, may also be indicated in the designs on particular garments, and, at least in the seventeenth century, women were also wont to express their literacy, and, in this appropriately indirect fashion, their learning (Cort, 1991).

According to Cort, landscapes depicted on kimonos may include objects which provided riddles and clues to connect the scene with specific references in classical literature, or they may boast embroidered Chinese characters, writing out, or simply hinting at Chinese or Japanese classical poetry, in both cases demonstrating a discreet form of communication about the wearer's education. Cort writes,

> the widespread value placed on literacy for women in the seventeenth century as a means of enhancing moral virtue and refining social graces emerged in the pattern of women's garments. That the patterns took on the form of riddles and allusions agreed with contemporary opinion that women should express their learning indirectly, by their behaviour, rather than blatantly. (ibid. 193)[2]

Warriors in Japan have worn a variety of helmets, many plated to protect their heads, but also often elaborately decorated with fierce figures and motifs, undoubtedly designed to strike fear into the enemy and display and bolster their own bravery. These helmets still form an important symbolic part of a small boy's informal education in modern Japan, for decorative versions of them are presented to families when sons are born, along with armour and other items symbolizing bravery and other masculine virtues. They are brought out every year and displayed in the best room for a period preceding the day dedicated to boys of the land. Girls are given a set of dolls, dressed in the elegant garments of the court, and usually adorned, with an elaborate retinue, in the finery of the wedding-day.

Another common form of headdress in Japan is simply made from a strip of cloth, or a towel, wound around the head and tied at the back. Useful for absorbing sweat which may otherwise drip and impede the progress of physical labour, it is symbolic, first of all, of making some kind of effort. On ceremonial occasions, such as festivals, or the building of a new house, the

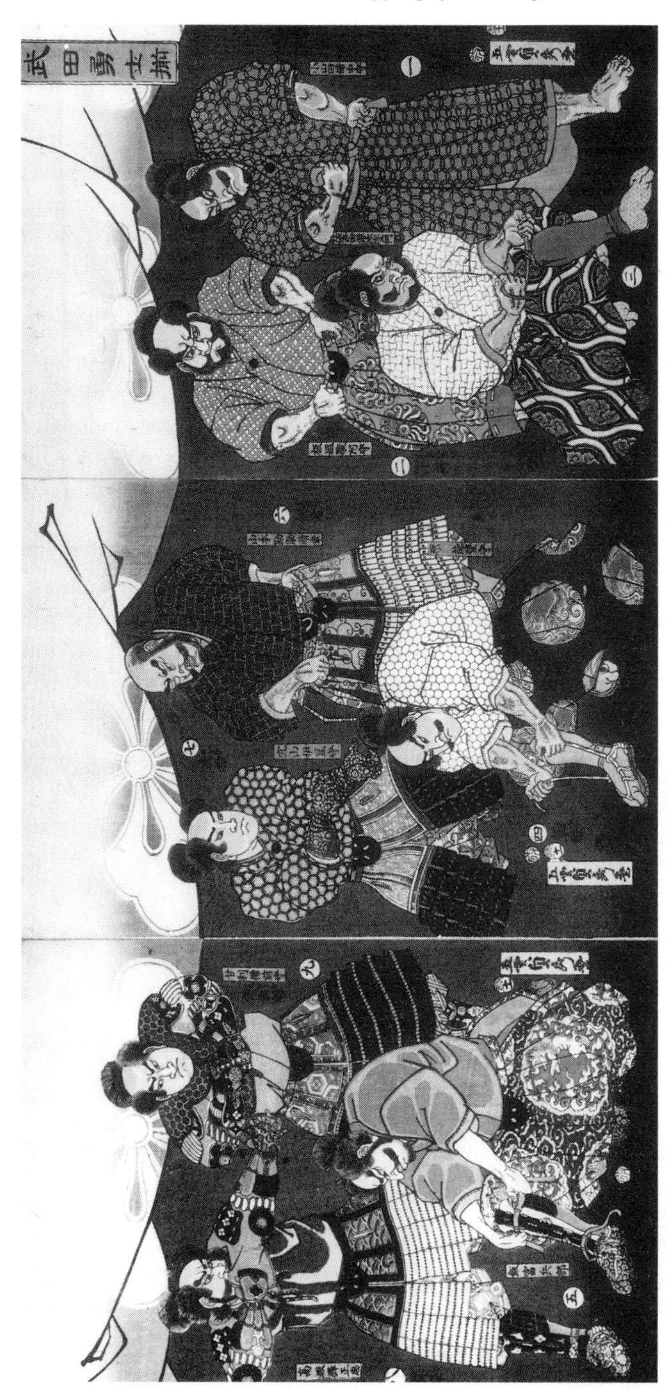

FIG. 4.8. This Sadahide (1853) print, *Nine stages of donning armour*, suggests that even this was a form of wrapping; by courtesy of the Board of Trustees of the Victoria & Albert Museum.

hachimaki, as it is called, may be dipped in a red dye to denote the festive nature of the effort, or it may be of different colours depending on the neighbourhoods participating in the festival (Fig. 6.4 and Plate XIX), each participant thus demonstrating their local allegiance to the wider gathering. More specific information may be written or stamped on this convenient location, one modern example being the way children sitting entrance examinations may wear *hachimaki* urging them to put great effort into passing.

This garment may be created from a towel like those discussed in some detail in Chapter 2, and the Japanese forbear of the object, the *tenugui*, has, according to Segawa (1979: 474), for long been worn on the head for various types of labour. Indeed, the *hachimaki* used for festivals would appear to be a specific example of the use of *tenugui*. A new one would be given at New Year to anyone who helped to pound rice to make *mochi* cakes, and the housewife who went out for the first water at New Year would also wear a new *tenugui*. In both cases, this would symbolize the sacred nature of the work being done—hence its use also as a *hachimaki* at festivals (ibid.).

There are, of course, many sacred garments worn by priests and other religious functionaries in Japan, and each of these could no doubt be fruitfully examined for symbolic meaning. Here we will limit ourselves to a single example, however, and consider the white robes donned by people making one of the many pilgrimage journeys available throughout Japan (Fig. 4.9). As mentioned above, white robes are also worn by brides, babies, and corpses,

FIG. 4.9. Pilgrims at the Kirihata Temple in Shikoku, no. 10 on the famous 88 site pilgrimage; courtesy Ian Reader.

and as such, they symbolize participation in a rite of passage, a symbolic death to one state ready for rebirth in a new one (cf. Reader, 1987: 140).

The donning of white robes is part of a rite of purification, sometimes also involving ritual washing, or even complete submersion in a river (Swanson, 1981: 63). The white symbolism here can be compared with the association of purity and sacredness with white *washi*, but it also has the significance of wiping the slate clean to start a new phase of life. According to Reader, the donning of these garments also nowadays helps people to experience a renewal or regeneration accompanying a sense of return to their spiritual roots, part of a nostalgic reaction to modern, industrialized Japan (1987: 140).

In reference to medieval pilgrimage, Grapard writes about the awe with which pilgrims returning from sacred places were regarded, so that people 'saluted them, made offerings, even tried to touch them', undoubtedly related to the fact that 'they bore elements of the sacred back into the profane world' (1982: 207; cf. Blacker, 1975: 100). This aspect of communication with the sacred suggests another parallel between the white garments of the pilgrim and the white fronds of the *gohei*, which, like the curly wood-shavings of the *inau*, were discussed in Chapter 2 as vehicles of spiritual power. The spiritual aspect of this 'magical link', as Kyburz (1988) called it, is perhaps weaker now, although still expressed through the purchase of *miyage* and *omamori*.

In Japanese drama, too—Nō, Kyōgen, and Kabuki—costumes and masks worn by the actors carry a multitude of meaning and significance. The costume itself will depict a character of a particular designation or occupation, details of style and colour indicating further distinctions of social status and inclination. Masks are used in Nō and these invariably belong to a predetermined set which represent various well known roles and characters. They are used with great care to express subtlety of meaning so that even a slight inclination of a mask can denote a change of mood.

Another powerful type of Japanese bodily wrapping, this time somewhat more permanent than many forms of attire, is the tattoo. Associated by many with gangsters and other men and women on the fringes of acceptable society, tattoos serve within these groups to identify the particular associations of their owners. The tattoos themselves can be extremely elaborate, expensive, and painful to acquire, and they thus demonstrate both the physical perseverance and the economic ability to acquire them. Depicting folk characters and brave heroes of Japan (Fig. 4.13), this form of adornment serves something of the same function as the samurai helmets, particularly in prison, and in view of the propensity of Japanese citizens to perform their ablutions in public bath houses. There would appear then to be something of a magical function, but the aesthetic qualities of the designs also play an important part in this example of bodily art (for further detail see e.g. Morita, 1966; Richie and Buruma, 1980).

A measure of the zeal with which dress of one sort or another is used in

FIG. 4.10. Theatrical Kunisada print (*c.*1810–15) showing black-clad stage-hands adjusting the scenery; © The Trustees of the National Museums of Scotland 1993. See also Pl. XIII.

Japanese society is to be found in the way Western clothes have been adopted since their first introduction only just over a century ago. Unlike many other parts of the world, where foreign clothes were only gradually incorporated into the more traditional wardrobes of native peoples, the Japanese began to sport a great variety of Western garments soon after they opened their ports to trade with the outside world. There were some early hiccups, such as a tendency to wear bowler hats with kimonos, and these continue through to the extraordinary slogans which appear on some modern T-shirts. On the whole, however, Japanese businessmen are as well turned out as any, and their

FIG. 4.11. Actors in Nō theatre are almost entirely wrapped; courtesy Japan Information and Cultural Centre.

women wear the most elegant of up-to-the-minute costumes.[3] Japanese fashion designers also appear among the most *avant garde* in the European world.

Uniforms have been very seriously adopted in the modern Japanese world. Apart from categories found everywhere, such as schoolchildren, nurses, and the armed forces, there is also a great variety of Japanese uniforms for functionaries of more commercial ventures. Those entering a large

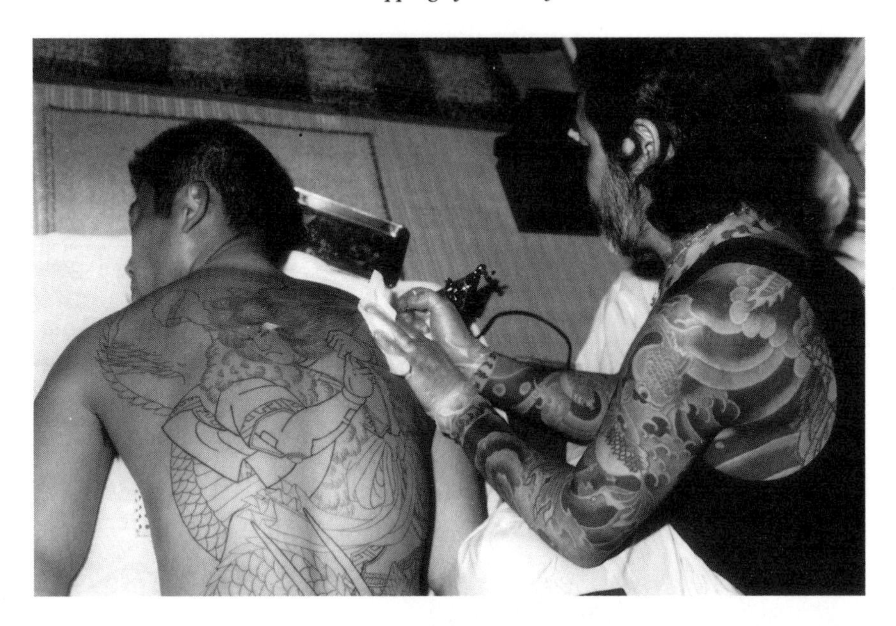

FIG. 4.12. A tattoo artist at work.

FIG. 4.13. Kuniyoshi print (1827–30), showing the legendary tattooed hero Rōshi Yensei; © The Trustees of the National Museums of Scotland 1993. See also Pl. XIV.

department store, for example, will usually be accorded a crisp greeting from a well-heeled young man or woman, situated just inside the door for that very purpose. Lift attendants (Fig. 4.14) wearing the same uniform utter information about floors in such stilted tones that they might almost be machines, but here is an example of the correlation between linguistic and bodily wrapping. In both cases, the individual human being is so well enveloped inside the company packaging that it is virtually impossible to break through this parcel to have any kind of interpersonal exchange.

This layer of wrapping also serves to protect the incumbent from such personal communication in an occupation potentially open to the possibly unwelcome advances of strangers, as was indicated for the use of formal and polite language in the previous chapter. Other functionaries whose language could almost be issuing from a machine include tour guides and bus-drivers who provide information about locations near the stops they make. In some cases these roles are provided by machines, but where a human being is carrying out a mundane and repetitive task like this, they will almost certainly

FIG. 4.14. Lift operator in a department store.

be neatly attired in the company uniform, usually down to a pair of clean, white gloves (Fig. 4.14), which provides a last symbolic separation from the public with whom they deal.

Despite this avid adoption of Western dress in Japan, traditional Japanese garments have by no means disappeared. The warmer kimonos are still worn in winter, usually by older women, but sometimes by men, too. More ceremonial garments are still worn for special ritual occasions such as New Year (Fig. 4.15), weddings, festivals, and children's celebrations of particular ages. The attainment of adulthood, at the age of 20, is a time when quite considerable sums of money are put into dressing daughters. Beautiful garments are purchased for these same daughters to take with them when they marry, and part of the betrothal gifts may be an expensive Japanese kimono, sash, and jacket.

Light summer kimonos called *yukatas* are worn outside on hot summer evenings, and many men will change into these when they return home from work in the evenings. They are provided, too, in hotels and inns, suggesting that many Japanese prefer the loose, comfortable traditional garments in which to relax, as well as for more formal, stiff attire. If this is the case, one

FIG. 4.15. A family dressed up for New Year's Day; courtesy Japan Information and Cultural Centre.

FIG. 4.16. (*Left*) A child formally dressed for a celebration courtesy Takahara.

FIG. 4.17. (*Right*) The same child, with her father, in a *yukata*; courtesy Takahara.

may wonder why the Japanese were so keen and complete in their adoption of Western clothes for the workplace and other modern settings such as cinemas, restaurants, and so forth. Some high-ranking Arabs and Indians have made no such concessions to the world outside their own culture, and they even travel to countries quite unsuited practically to accommodate their splendid floor-length gowns. In the last section we will turn to examine some of the implications of this cultural variation.

Intercultural Confusion in Bodily Wrapping

In the world of the late twentieth century, success for nations is measured in terms of economic achievement. Japan has now streaked ahead of her staunchest Western rivals, sparking off innumerable studies of her methods to gain this extraordinary accomplishment. In the Middle East, on the other hand, some countries have become remarkably wealthy, while others have remained poor in global terms, clinging nevertheless to other principles, perhaps religious, perhaps cultural.[4] It might seem at first sight that clothes can have little to do with this, then, since representatives of rich Arab nations continue to wear the robes which distinguish them in their own societies, whereas the Japanese have taken great pains to fit in elsewhere. I would like to argue here, however, that clothes have in each case plausibly had quite a part to play.

The discrepancies in the Middle East are of course complex, and quite clearly related to good fortune in relative access to natural resources, such as oil, which are in constant demand in the world market. The Japanese, on the other hand, have few natural resources to play with in world monopoly, and the secret of their success is sought rather in terms of human potential. Their clothes have attracted little attention, except perhaps for the occasional giggle about minor mistakes in their use of ties and badges, or howlers on their T-shirts, but this may even have been one of their strengths. A man in an unremarkable suit, trying hard to conform to local custom and fit in with his surroundings, especially if he fails occasionally, appears to pose little in the way of a threat. Although few Japanese would enjoy being described as ridiculous, it may paradoxically be this slightly ridiculous air which has helped them at times to gain access to information vital to their economic development.

For suits are what Western people serious about their business have now been wearing for so long that they seem normal. A person wearing a suit is a person whom one can address, a person with whom business can be conducted, even if an interpreter is required to translate the actual words spoken into the language appropriate to the country concerned. It is of no great interest to a Western representative if the person they are facing has quite different thought patterns. This is something which may never enter the head of someone on their home territory. What are different thought patterns anyway? They are something weird that anthropologists and philosophers talk about, if they have been heard of at all, certainly nothing which need interfere with the balance of trade, or the state of the stock market.

Furthermore, if the person in the suit seems at a disadvantage, perhaps because they obviously fail to understand some of the niceties of etiquette, or because their language is far from perfect, there appears to be no great need to worry about their potential as rivals. If they ask to see the latest machines in the factory, or the plans for new developments, why not encourage their interest? After all, they asked politely enough, and they can't possibly really understand them. How nice that this foreign gentleman should show such interest in my area of specialist knowledge. He did bring a splendid gift for my wife, and he has invited us all to dinner at the Japanese restaurant. If he comes offering some further benefit, like a free trip to Japan, or badly needed funds for the latest research project...

The case has perhaps been exaggerated, and the *naïveté* of the host overemphasized, but let us press the argument further and consider the case of uniform. Uniform serves a number of purposes. First of all, it identifies the wearer as a member of a particular organization, usually also assigning that individual an occupation and a role in society. Within the organization, and in some cases more widely too, it may provide other members with further information about the person's position in an internal hierarchy. Nurses in a hospital, for example, wear different colours depending on their rank, and those with the same coloured dresses may wear different belts or cuffs to indicate further, finer distinctions between them. Members of the armed forces, or the police, display their ranking in similar ways, in stripes or stars. All this information is clear if you understand the system.

Let us now turn, briefly, to imagine ourselves in a society which has never seen a uniform, nor indeed a suit. It is, of course, quite impossible to imagine what kind of an impression such garments might make on an almost naked native, but the words 'almost naked native' are quite evocative in the reverse case: 'almost naked'—quite naughty, especially for those brought up in the biblical tradition of Adam and Eve, certainly inappropriate for the city; 'native'—well, the literal term in itself is innocuous enough, but there are undertones of great variety—dangerous, strange, uncivilized, perhaps romantic. It doesn't tell us much, and the suit and the uniform merely tell the nearly naked native that the visitor has developed some technical skills in the production of bodily wrapping, little more.

It was the custom of explorers and colonial administrators in many parts of the world to try to ascertain who, amongst a native people they might meet, was in charge. Who was 'the chief'? Dealings with 'the chief' would be best, it was thought, because they would be likely to have some control over the other people—or so it was thought. In certain societies the system worked well, for there was a hierarchical structure involving control and representation. In others, there was no such thing, and attempts to designate a member of a particular tribe as the leader to deal with the outside world undermined any leadership the individual may have had to that point. In parts of Latin America, local 'leaders' were ridiculed by their own people because they agreed to wear uniform provided by outsiders trying to administer the area, and amongst the Trio Indians of Surinam, Peter Rivière reports (personal communication) that a local idiot was even pushed forward by the others to take on this role (Figs. 4.18 and 4.19).

Uniform simply made such individuals look preposterous to their own people, and reactions were strong. In such societies, leaders lead by exemplifying the values of the society, by being good hunters, by being athletic, and, as elsewhere, by their rhetoric. It is important not to appear too

FIG. 4.18. (*Left*) Trio Indian (Surinam) in uniform provided by administration; courtesy Peter Rivière.

FIG. 4.19. (*Right*) Trio Indian in habitual wear; courtesy Peter Rivière.

dominant, however, and it may be quite taboo to issue orders. Leaders literally lead. They demonstrate their leadership by getting on with what needs to be done, and others, if they respect them, will follow. If they don't, then the leader has lost his influence, and his power, if what he had can be called power. It's a case of rather exemplary democracy, if nothing else, and only a fool would go off and get dressed up in the clothes of the outsiders.[5] A fool by their terms, that is, for the Japanese are no fools.

In the modern period of Japanese history, the Japanese have become involved in several wars. They had fought many wars of their own in the past, with quite different rules, and wearing quite different clothes. Before they became involved in Western wars, they took great trouble to develop Western technology, guns, gunpowder, ships, tanks, and aeroplanes. Moreover, they kitted themselves out in perfect replicas of Western clothes. The Japanese navy, for example, is modelled on the British navy, and they still look uncannily familiar in charge of their ships. At meetings to discuss treaties and formulate alliances, they turned out immaculately dressed. To all intents and purposes, they were real people who would obey the accepted rules of the game, the game of war. In practice, of course, when it came to real battles, they didn't obey the same rules, and they shocked the world with their behaviour.

In previous encounters between Western nations and strange, foreign peoples with whom they fought, it had been the Western nations who committed the atrocities. At least, that's the view we are beginning to realize now, as we look back on some of the terrible massacres we inflicted on 'Red Indians', 'Black Zulus', and 'Brown Savages'. Of course, these were people without technological advantages, but that only makes it worse. Almost certainly the justification for those personally involved at the time included some notion that they were not dealing with real people, anyway, not people like ourselves. The colour of their skins may well have made a difference, but I suggest, too, that their bodily wrapping played an important part—or the apparent lack of it. Feathers, paint, and penis sheathes hardly compare with uniforms, and the rules, already quite well established within intra-European encounters, could easily be cast aside under such circumstances.

To support this argument further, let us turn our attention to other parts of the world where 'natives' were already more impressively 'wrapped'—from our point of view. The British and other European colonists managed to communicate rather well with certain African chiefs, particularly those with 'kingdoms', as we came to call them, with readily identifiable kings and princes. These characters, on the whole, wore splendid garments, perhaps flowing robes, jewellery, and other evidence of superior wealth. They were also often enough surrounded by other individuals, again dressed in special garments, a little less splendid than those of the supreme leader, but nevertheless indicating that they too occupied positions of rank.

Just like Europeans, these societies used clothes to demonstrate their social system to the world at large, to distance their chiefs and leaders from the ordinary folk. At a time when kings and princes in our own countries were walking galleries of art and wealth, gathered by their subjects from all corners of the known world, this was a system with which we could identify. Of course, technology played a part in forming the impression of relative civilization, but our ambassadors also had a fund of appropriate ways to behave in such circumstances, and in many cases, these were suitable for local use.

It may seem strange now, but Elenore Smith Bowen's (1954) description of dressing for dinner during fieldwork alone in an African village, and her insistence that her servants use the proper crockery and cutlery for this occasion, brought home the force of this point for me. Today's anthropologists are generally much keener to adapt to local circumstances, but the power of such signs of 'civilization' was undoubtedly great for other Europeans abroad. Indeed, our sumptuous symbols of regalia may have played no small part in the way we were able, with relatively little aggression, to move in and successfully rule whole areas of people. India is a case in point, likewise several African states. Small wonder the British diplomatic service still provides its employees with a dress allowance for their wives.

Two kingdoms, or in this case, empires, were impervious to our approaches, however, and we never managed quite to impress them enough, although we were undoubtedly impressed enough by them to make the overtures. These were China and Japan. Undoubtedly, one of the reasons they impressed us was because they also knew how to wrap the bodies with which they were born. We are still impressed by traditional Chinese and Japanese garments, imperial and otherwise, and we hang them in our museums. What would have happened if we had tried to emulate them, as the Chinese and Japanese have now done in the international arena, we will never know. We do know that T. E. Lawrence made a great impression in Arabia, particularly after he adopted the robes of the desert, but he was regarded as a fanatic by his countrymen.

In fact, these garments are highly significant to the people who wear them, as Lawrence no doubt appreciated, and once robed in that way, further communication becomes possible which is quite esoteric. Among the Tuareg, a people described as 'men of few words' who 'express themselves only by hints and understatements' (Casajus, 1985: 72), the veil the men wear after puberty is a particularly interesting case in point. First of all, there are various possible ways to wear it, depending on whether one wants to appear 'reserved and modest', 'elegant and recherché', 'haughty' or 'detached', each an attitude appropriate for particular occasions with particular groups of people, which may be changed as circumstances change or as the 'tone of a relationship shifts' (Murphy, 1964: 1266). Moreover, because the eyes and sometimes the nose are the only part of the face left visible (Fig. 4.20), a great deal is also communicated through 'the position of the eyelids, the lines and wrinkles of the eyes and nose, the set of the body and the tone of the voice' (Murphy, 1964: 1265).

The overall importance of the veil is evident in the fact that even when these people adopt Western clothes—in one case, according to Murphy, yellow plastic sandals, blue shorts, and a checked sports shirt—they are quite likely to continue wearing the veil and turban with them (ibid. 1264). Murphy's view of the significance of the veil is particularly interesting here because, according to him, it plays a role of social distancing which he feels is comparable with that set in other societies by joking, respect, or avoidance behaviour. It would thus seem that this type of bodily wrapping has a role not dissimilar to that outlined in the previous chapter for the linguistic wrapping known as *keigo*.

Tuareg women, on the other hand, unlike other Muslim women who live

FIG. 4.20. Tuareg man; courtesy Nairn/ Disappearing World/ Hutchison Library.

around them, do not veil themselves, and according to Murphy, they have privileges unknown to these neighbours. They are neither secluded, nor diffident about expressing their opinions, they do not value pre-marital chastity, nor do they defend monogamy. Moreover, they apparently have male friends other than their husbands and they can secure a divorce merely by asking for it (ibid. 1262 n.). This life is described somewhat in opposition to the expectations for Muslim women who do veil themselves, and it would seem quite possible that these differences in bodily attire reflect differences at other levels in the social structure.

A more recent study of the Kel Ewey Tuareg of north-eastern Niger (Rasmussen, 1991) makes explicit comparisons between the men's veils and headscarves worn by the women during their childbearing years. Rasmussen finds several similarities between the two forms of headgear, interpreting a playful use of both as 'bringing about a compromise between personal goals and social constraints' (ibid. 102). In this respect she likens them to masks in other African societies, especially when they are used during possession rites, emphasizing that in both cases Westerners mistake their role as one of disguise whereas from an African viewpoint they are 'transformers of the wearer's identity' (ibid. 103).

Rasmussen's discussion of the way the significance of the women's headscarf varies with context is also reminiscent of the use of *keigo*. In some contexts, it signifies subordination and respect, in others high status and prestige. It may be adjusted to express modesty and shyness towards strangers, or, on the contrary, to tease and to flirt (ibid. 115). All these examples could still be applied as possibilities for the skilful use of *keigo*, and parallel ones could no doubt be suggested for ways of wearing kimonos when this was the usual form of attire.

Other factors are obviously important, over and perhaps above the bodily wrapping we have been considering in this chapter. In the following chapter we will turn to examine another important form of 'wrapping', including a brief examination of another parallel in the Muslim world, but I would like to finish with a final demonstration that we may do well to take care in the assumptions we base on this criteria. For the most part, we dismiss as 'savages' people who display too much of the bodies they were born with—unless they are in the bath, or sitting on a beach. Recently, however, a group of South American Indians, apparently insistent that they are happily garbed in feathers and beads, have successfully found another means to impress not only the Brazilian government, but the media of the world more widely.

The Kayapo (Fig. 4.21) have become excellent computerized capitalists, defending their land by charging their government highly for the privilege to use and develop it. They have acquired considerable wealth in bank accounts and plastic, media well understood by the outside world, and they take care to record and replay all verbal agreements, to avoid the exploitation suffered by

FIG. 4.21. Kayapo men with Granada Television crew equipment; courtesy Mike Beckham/Disappearing World/Hutchison Library. See also Pl. XV.

many of their near compatriots in the struggle to maintain an existence which they value for quite different reasons (Disappearing World film: *The Kayapo*). They have managed to find a means of communication without sacrificing their own modes of attire, as have the oil barons of the Middle East, and they may have brought a modicum of respect for the near naked human form, though it may for some time require an anthropologist, a 'green' film-maker, or someone seen by the wider world as a freak, to find much value in the world they are concerned to maintain.

5

The Wrapping of Space

The previous chapter addressed a particular form of material wrapping, namely that used to wrap the body, a form which is portable, and which has immediate impact, a form which may be changed with relative ease. This chapter will turn to examine a form of material wrapping which can be much more permanent, much more expensive and time-consuming to create, and ultimately impressive in a much more profound way. This is the 'wrapping' of space. Through the wrapping of space, a people becomes a civilization, and the wrapping which they create remains to impress the world long after the people themselves have disappeared.[1] This is the legacy left for history and archaeology, ultimately this is the wrapping which has survived most successfully at least until the invention of modern technology.

It may seem a little far-fetched to call buildings wrapping, although as two architects have recently pointed out, they are different from other artefacts in that they 'create and order the empty volumes of space' and 'this ordering of space . . . is the purpose of building, not the physical object itself' (Hillier and Hanson, 1984: 1). The need to explain buildings in this way perhaps betrays a Western bias, for another author has commented, 'For many Westerners a house is a thing, an object . . . But to the Japanese it is a context—or rather a shifting set of smaller contexts within a larger one . . . not a box with openings, as in the West, but a space-moulding system' (Greenbie, 1988: 14). Coaldrake feels that both Western architects and Japanese architects trained in the West have generally misrepresented Japanese ideas of 'space'. He writes, 'in Japanese architecture what is seen is frequently not there, and far more is present than is actually seen' (1988: 113).

I hope eventually to show that buildings in general offer a rather special type of wrapping, with some almost universal features, but this chapter will start with the Japanese case, which is quite plausibly brought into the same category as other Japanese forms of wrapping. It is possible that there is a Japanese conception of space which lends itself to this form of description. The reader will perhaps be able to make that judgement as the case unfolds. No one will dispute that space is profoundly transformed by buildings, both from the inside and the outside of them, and it is true too that people are judged by the space in which they live, just as they are judged by the clothes they wear and the cars they drive. Let us examine some of the other possible meanings of spatial wrapping.

Japanese Spatial Wrapping

The Japanese wrapping of space reflects the use of wrapping in other ways in that it may involve several layers, each having some special significance, and the whole designed to create a particular effect. These principles are found in domestic areas such as houses, gardens, and palaces, in public buildings such as town halls and ministries, and they are particularly marked in sacred locations such as shrines and temples. The use of the space thus enclosed reflects the design, and certain categories of people will be permitted to penetrate the space only as far as their status and position allows. Again the wrapping reflects degrees of distance and formality and the hierarchical structure of the people using the space.

To start with ordinary domestic space, a Japanese house is usually designed in such a way that it can be penetrated to various different degrees. First of all there is a porch, which may be entered with very little ceremony by anyone who cares to call at the door. It is here that bills will be paid, messages left, and other minor business negotiated. The only ritual to be observed by those who enter is to call out a greeting, and in the country the outer door may be opened without even knocking or pressing a bell. This is unwrapping the outermost layer of domestic space, but the level of communication is usually fairly distant.

To move further into a Japanese home, various small rites must be observed. First of all, an invitation must be issued from inside for the visitor to 'come up' into the house. The visitor must then remove his or her footwear, which is left behind in the porch, and phrases are uttered both from inside and out as the person moves up into the inside hall. If the visit is to be of more than a short duration, slippers will probably be offered to the outsider to wear within the house. A fairly formal or official visit may nowadays take place in a Western room with a table and chairs, and this is as far as the outsider will penetrate. Insiders will come and go through a door, traditionally a sliding door, and glimpses of the interior may be snatched, but the guest will go no further.

In a more traditional Japanese house, or a modern one where a Japanese ambience is preferred, a guest may be received in a *tatami* matted room, and to enter one of these the visitor would need to remove the slippers again. In some country houses, there is a room beside the porch called the *ozen*, or 'front room', where mundane business would be transacted, and this is usually floored with *tatami*. A more elaborate occasion, perhaps for some special ritual, will be held in front of a raised dais known as a *tokonoma*, which is located beside the principal post of the house construction, in a room known as the *zashiki*, the most formal room of the house (Fig. 2.9). In the higher class country house, there would be yet another room between the front room and this ceremonial one (Hendry, 1981*a*: 83 illustrates this).

The *zashiki* could perhaps be termed the ceremonial centre of the house, but this room will still be on the outside, or visitors' half of the house, which is usually clearly separated from the insiders' more intimate half[2] (Bachnik, 1986: 64 illustrates this). Insiders will still move in and out, again through sliding doors, and refreshments will probably be brought from the inner sanctum. This latter is referred to as the *oku* of the house, the 'heart' or 'interior', although also having connotations of 'depth'. The same character is used in the compound for 'wife', namely *okusan* (cf. Pezeu-Masabuau, 1981: 51), and the character itself portrays rice, the main means of subsistence, entirely enclosed in other logographs.

To enter the inner sanctum requires no further ritual, but rather a degree of intimacy between the householder and the visitor. This will be accorded to different people depending on household practice and the custom of the region where the house is located. In a sophisticated household of the upper classes, few visitors would be allowed beyond the formal areas; in a country house, a relative stranger may be invited into the inner family area once the formalities of greeting have been dispensed with. The inner part of the house is usually much more comfortable, often literally much warmer, and it is only here that it would be possible to feel relaxed and at ease.

It is not necessarily that there is a centre, although in the traditional structure this could possibly have been the hearth, and one Japanese architect feels that the raised dais known as the *tokonoma* is the centre (Yoshida, 1955: 10). However, the house is designed in such a way that the inner areas appear quite distant from the outer ones, and the sliding doors appear to wrap the inner areas and protect them from the outside (Fig. 5.1). Just as with language, layers of polite formality conceal (and occasionally reveal) an inner sanctum, and the nearer the outside one finds oneself, the more formal is the expected behaviour. The same applies to the formality of garments, the food consumed, and the gifts exchanged.[3]

Some of the effect of depth in the Japanese house design is in fact an illusion created by the combination of paper screens and sliding doors, and this art is found also in gardens. There is some notion here of 'creasing' or layering of space to create an impression of depth and mystery (Fig. 5.2). The Japanese architect Fumihiko Maki (1978, 1979*a* and *b*), regards this creation of *oku* as a distinctive feature of Japanese architecture which he contrasts with the 'center-demarcation' of Western cultures. Western architects have also commented on the effect. The Hungarian architect Botand Bognar, for example, comments on the 'thin wall systems layering the space in a multiple way,' and notes that 'an outer space always seems to envelope another one inside' (1985: 60). The suggestion of an innermost 'core' is purely theoretical, he argues however, 'in a sense an invisible, unattainable zone', and he goes on to add, 'the Japanese have always been able to give an illusion of depth to

FIG. 5.1. Japanese architecture appears to wrap layers of space.

FIG. 5.2. The layering effect is found in Japanese gardens too; this one is at the Hotel New Otani, Tokyo. See also Pl. XVI.

spatial compositions, regardless of their actual, usually small or shallow, dimensions' (ibid.; see Fig. 5.3).

Spatial wrapping in castles and palaces follows the same principles, but in a more elaborate form, and the surroundings of the building are usually constructed to add extra layers, thus offering protection to the higher ranking residents from the ordinary people without (as well, of course, as from their enemies). From outside looking in, a perception of distance and formality helps to create an aura of power[4] (Fig. 5.4). As Coaldrake writes of late sixteenth- and early seventeenth-century Japan in the introduction to his translation of Hinago Motoo's *Japanese Castles*, 'The castle also became the

FIG. 5.3. Kunisada II print (1867) showing layering of space in a courtesan house; © The Trustees of the National Museums of Scotland 1993. See also Pl. XVII.

FIG. 5.4. Castles 'create an aura of power': Himeji castle; courtesy Japan Information and Cultural Centre.

FIG. 5.5. Sometimes castles and palaces are so well 'wrapped' they are hidden behind the layers: Aizu castle exterior.

proud statement of the accomplishments and authority of the samurai class, the palatial residence of national rulers, and a glittering center of court ceremony and patronage of the arts' (Coaldrake, 1986: 18).

The Imperial Palace in Tokyo is to this day an excellent example of the principles expressed here, although the associated power is more complex (to be discussed again in Chapters 6 and 8). In this case the outer gardens and moats obscure almost entirely the buildings which they surround, and the astronomical value of the land they occupy cannot help but instil some awe in any member of the ordinary population who pauses to consider it. In fact, when I visited the Imperial Household Agency, enclosed within these grounds, to acquire the photograph of the imperial wedding for Fig. 4.3, the taxi-driver I asked to take me there was at first aghast. He warmed to the idea though, and waited patiently while I carried out the various rites required to penetrate the layers of protection.

Having made a request for the photograph by telephone, which involved explaining many details of my purpose, I had to go in person to make the request again, with the same degree of detail. I then had to put all the same information in writing, and wait for a few days for an answer. Each time I went there I had to make an appointment by telephone, hand my calling-card in at the outer gate, where I received a badge and permission to pass through. This I had to show at the next gate, after which I could enter the administrative building. Strangely, at this point I had to find my own way, but I understand this is not the case in the palace itself. I had only entered an outer layer of the imperial wrapping!

Castles were often set upon imposing hills, with huge stones forming the basis of their foundation construction, so that although they were visible from afar as symbols of the power and authority of their occupants, they were not easily approachable (cf. Hirai, 1973: 10, 15). It would be virtually impossible for ordinary people to enter such buildings when they were occupied by their original military leaders, but now they are often open to the public to examine. Again, there are inner and outer halls and apartments, and the seat of the lord, or his representative, is usually in an inner chamber well protected with lesser chambers for lesser visitors, and chambers for his advisers on an inner side. According to Japanese literature about the courtly period, much use was also made of screens, as a form of spatial wrapping. The point is illustrated in the following quotation about the enthronement room in the Imperial Palace in Kyoto:

> It is approached from the front through two verandas; the inner one . . . is marked off by rows of round columns. . . . Between the columns of the outer row hang vertical *shitomido*, or paper-covered wooden lattices . . . A less definite degree of separation is attained with the use of *sudare*, thin, roll-up bamboo shades, suspended just outside the *shitomido* and also between the columns of the inner row. (Alex, 1963: 27)

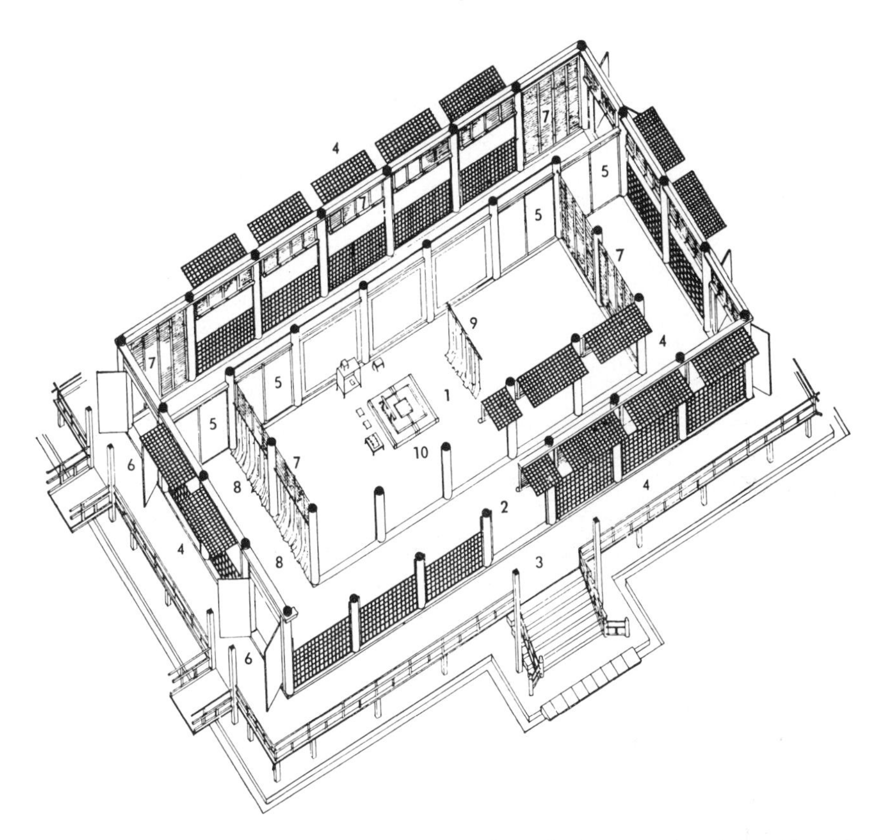

FIG. 5.6. Conjunctured interior arrangement of the *shinden*, reproduced from Yoshida (1955: 25): 1 *Moya*, principal room; 2 *Hisashi*, verandah; 3 *Sunoko*, open verandah; 4 *Shitomido*, wooden flap; 5 sliding door; 6 hinged doors; 7 *Sudare*, bamboo curtain; 8 cloth curtain; 9 *Kichō*, movable stand hung with curtains; 10 *Tatami* mat with cushion.

This palace has been continually reconstructed since it was originally designed at the end of the eighth century, and its layout provided a model for a classic domestic style known as *shinden-zukuri* which was developed for aristocratic dwellings between the tenth and twelfth centuries (Yoshida, 1955: 21). Here there is one large central room, surrounded by a closed verandah, itself surrounded by a smaller, open verandah, which was very often finished with a small fence. The large central space was divided up for particular occasions with the use of screens, or portable stands hung with silk curtains (Fig. 5.6), and the *fusuma* sliding doors are said to have developed from these screens (Yoshida, 1955: 23). In general, the notion of surrounding domestic space with one or more layers of verandah has persisted through to the present day, and illustrates very well the propensity for 'wrapping space' in Japanese design (Fig. 5.7).

FIG. 5.7. A verandah is a common feature in Japanese houses.

In public buildings, the wrapping principles are rather different, but a couple of examples will suffice to illustrate that they exist. In town and city halls, for example, there is usually no place to remove your shoes, and it is possible to walk in and penetrate quite far without any form of ritual whatsoever. It is common, however, for there to be a series of glass cases near the entrance displaying some of the objects for which the area is known. These may be examples of crafts, or models of some industrial activities, and an outside visitor curious about the region would need to go no further to get a good idea of the activities of the location. Indeed, the building itself represents the expense and effort the local council has thought worth portraying.

Notices displayed in the inner hall will also answer some of the queries for which local visitors may have made the journey, and these need go no further. People with specific business to negotiate will be directed to the appropriate department, and those commonly visited will probably be fairly accessible, perhaps 'wrapped' only by a counter. More unusual business may entail going upstairs, entering a closed office, and perhaps passing a group of people working in the department concerned. The heart of local politics, the debating chamber of the town or city council, is likely to be much further within, perhaps a room with no outside aspect at all. To wield power within a particular construction, one needs to know how to penetrate the various layers of wrapping.

Another example of the wrapping of a public building is a specific example of a different order. This is the Ministry of Justice, a building which must be visited by foreigners in Tokyo seeking to extend their official stay in Japan. The building stands at the side of the Imperial Moat, in a district populated largely with the imposing structures of government ministries. This particular one looks most splendid to a Western eye, for it is apparently built of brick, and its style is eminently European. It seems quite appropriate, then, for the purpose mentioned above, but this impression is deceptive, for the brick face is purely a façade, and the inner structure is of reinforced concrete like most of the other buildings around it. As I have already pointed out elsewhere, the Western face is quite in keeping with the way the Japanese legal system is itself wrapped in Western packaging (Hendry, 1987: ch. 12). It appears, from the outside, to operate like a Western system of courts and code, but in practice its operating principles are very different. In this way, many times, Japan satisfied her Western critics that she was no barbarous nation, but without making too many concessions to Western ways of doing things.

Maki's notion of wrapping as a characteristic of the Japanese use of space extends also to the design of villages and cities. An ideal type for the former is to nestle, surrounded by rice fields, against a backdrop of mountain scenery, at the foot of which will be situated the village shrine. Further away, up into the more remote areas of the mountain, will be an inner shrine, the *oku-sha*, he argues (Fig. 5.8). This he contrasts with the European pattern of clustering houses around a central area, usually featuring a church, with a spire (1978; 1979*a* and *b*). Cities, too, exhibit a basic difference of this sort, according to Maki, who characterizes the Japanese case as a process of 'inner space-envelopment', as opposed to the 'center-demarcation' found elsewhere (1979*b*: 102).

Roland Barthes's (1982) comments on the 'empty' nature of the city centre in Tokyo may be compared with Maki's contrast here, but this contrast is articulated in a somewhat different manner by Augustin Berque (1990), the French geographer, in his comparison of French and Japanese cities. He argues that the quality of the urban/urbane, in behaviour as well as in townscapes, is founded on different grounds in France and Japan. Whereas in the former it is based on the political relations between men, in Japan it is based instead on the de-politicizing relationship of man with nature. His elaboration of this point refers to the grounds of the Imperial Palace—Barthes's 'empty center'—as a 'forbidden forest in the heart of Tokyo'. 'Forbidden' it may be, but the notion of wrapping of such space lends much more significance to this extraordinary city centre.

There is, of course, a palace in the centre of this area, and the notion of an 'empty center' is created by Barthes in reference to the facts that this palace is concealed by foliage, the emperor is 'never seen', or 'literally, by no one knows who', and the traffic must constantly make detours around it (1982:

FIG. 5.8. Eisen print (1843–5) illustrating the 'wrapping' of a mountain shrine; by courtesy of the Board of Trustees of the Victoria & Albert Museum.

30).[5] In fact, contrary to Barthes's expectation, I think the place does 'irradiate power' of a certain sort, and the problem may lie in our Western propensity to want always to be unwrapping, deconstructing, seeing the objects at the centre of things. Bognar, too, comments about Japanese cities that 'if we try to lift the veils wrapping them in endlessly juxtaposed layers, surprisingly . . . [they] become "empty"' (1985: 67). Undoubtedly, what we need to do, is to learn to value the wrapping, as well as the wrapped, and seek the meaning they together convey.

Wrapping of the Sacred in Japan

Spatial wrapping in Japan is particularly evident in the sacred realms of shrines and temples. Even within the domestic sphere, there is usually a sacred space to be found inside the Buddhist altar, where the ancestors of the house are remembered. In older houses, particularly the larger ones, this Buddhist altar often has its own room, which may also be the place where the oldest generation sleeps. Within this room (or perhaps the *zashiki* if there is no special room), the altar is sometimes enclosed again within a cupboard. The front of the altar itself will also have doors, and the tablets representing the ancestors are further enclosed in a special container. The doors are usually opened for a special occasion, but a prayer or greeting to the ancestors is usually preceded by a ritual approach including the lighting of incense, the ringing of a bell, and the presentation of some form of offering.

Houses very often have Shinto shrines as well as Buddhist altars, but these tend to have an aspect open to the outside world, although positioned high up in a room. They are in the shape of a building, with an inside hidden away, but talismans and amulets purchased at larger shrines are just laid on the shelf rather than being stored away inside. These shrines are wrapped up during periods of pollution, however. If a member of the household dies, for example, the house is officially in mourning and a piece of paper is pasted up over the Shinto shrine to protect it from the polluting influence of death. Members of the household should also refrain from visiting community shrines at such a time.

The large Shinto shrines actually illustrate the wrapping principle very well, for they enclose their most sacred area with several layers of space which becomes increasingly sacred as an approach is made from the outside world (Fig. 5.9). The first marker is usually a *torii* or archway, and large shrines may have two or even three of these, the most distant one located quite a walk away. Other objects, such as stone pillars and statues, mark out the path which gradually approaches the shrine buildings. Most visitors will stop at the entrance to these buildings to ring the bell, offer a coin, and say a prayer (Fig. 5.10), although some special rites will be held inside the buildings (cf. Coaldrake, 1988: 116).

FIG. 5.9. The Meiji shrine in Tokyo is well 'wrapped', though the outer *torii* are out of sight; courtesy Japan Information and Cultural Centre.

FIG. 5.10. Worship often takes place outside the shrine building; courtesy Japan National Tourist Office.

Quite mundane activities may take place within the outer *torii* of a Shinto shrine. This is where itinerant vendors set up their stalls on festival days, for example, and the shrine in the village in Kyushu where I worked located the children's swings permanently inside the *torii*, apparently in the hope of receiving the protection of the deity. Many larger shrines have places for visitors to wash their hands and rinse out their mouths as an act of purification to be performed before proceeding towards the more sacred areas (Fig. 7.1). If the buildings are to be entered, the further purificatory rite of removing shoes must be observed.

Mitsuo Inoue (1985) has examined in some detail the development of the structure of space surrounding shrines and temples in Japan, and he traces the influence of Buddhist buildings on the layout of Shinto shrines through an examination of ancient texts and literature. The development of the *kairo*, a fence surrounding the area in which the building stands, is examined as a marker separating sacred and profane areas, and special attention is given to the gate which allows communication between the two. It was here that people would worship, and rites which take place at modern *torii* may reflect this practice. The imperial shrine at Ise, rebuilt regularly to preserve the ancient structure, is still ringed by no fewer than five fences (Fig. 5.11), and it is at

FIG. 5.11. The Ise shrine and one of its fences; courtesy Japan National Tourist Office.

one of the gates, the inner Tamagaki, that the imperial family worships, according to Inoue (1985: 43–4).

At the apparent heart of a Shinto shrine is housed a 'sacred body', a holy object, which represents the presence of the deity (*kami*), like the tablets in a Buddhist household altar represent the ancestors. This is usually wrapped and almost never sees the light of day. It is not in fact the object itself which is of interest, but the *kami* that it symbolizes (Ono, 1962: 21). The object is apparently often a mirror (ibid. 23), one of the three imperial regalia said to have been presented by the founding ancestress Amaterasu to her son who became the first emperor. This is an object with significant roles in the mythology of Japan's creation, but the idea that it reflects back what is outside would also seem to be full of significance.

In this context, it helps to point up the importance of the surrounding wrapping, and it is particularly interesting that worship usually takes place at one of the outer layers. It is again not necessary to seek the centre at all. As Bognar has pointed out,

> the Ise shrine stands as a huge symbolic object rather than a building with interior space, and as such its exterior space-organizing role increases in importance. Shrine buildings, starting with Ise, are in effect not meant to be spaces to enter, but rather remote places to approach and arrive at. (1985: 44)

Here is another reminder of the need to adjust our own culture-bound inclinations to unwrap, rather than focus on the wrapping itself.

These sacred spots are in fact thought to be points of contact, where deities can be summoned and communication take place. They are *yorishiro* like the staff discussed in Chapter 2. They may be natural objects, such as a stone or a tree, and in some cases the deity is thought to be present in a mountain, usually approached by more than one shrine to lead the worshipper towards the sacred area. In the past, a sacred site could just be open spaces, kept clean and clear, but they would be marked off by some form of wrapping (Speidel and Duff-Cooper, 1990: 295–7; Grapard, 1982: 196–200).

As mentioned in Chapter 2, sacred space is marked off with the same materials as are used in the wrapping of gifts, chiefly paper and straw. A *torii* at the entrance to a shrine will sometimes be hung with a plaited straw rope (Fig. 2.10), for example, and a thin version of the same may be used to mark out an altar for the ground-breaking ceremony on a building site. In some areas, a whole neighbourhood will be marked out in such a way in preparation for a local festival, and most houses in Japan hang up a straw construction on their front door, and very often on their car, too, at New Year. The straw rope worn by sumo wrestlers has sacred connotations too. Paper is used for similar purposes, as was discussed in Chapter 2. Straw ropes are often decorated with hanging paper streamers, and a priest will be seen cutting paper in preparation for many rites.

Pl. I. The receipt of a nicely wrapped gift may give pleasure.

Pl. II. Gift-wrapped steaks—a popular midsummer present. (*See also Fig. 1.4.*)

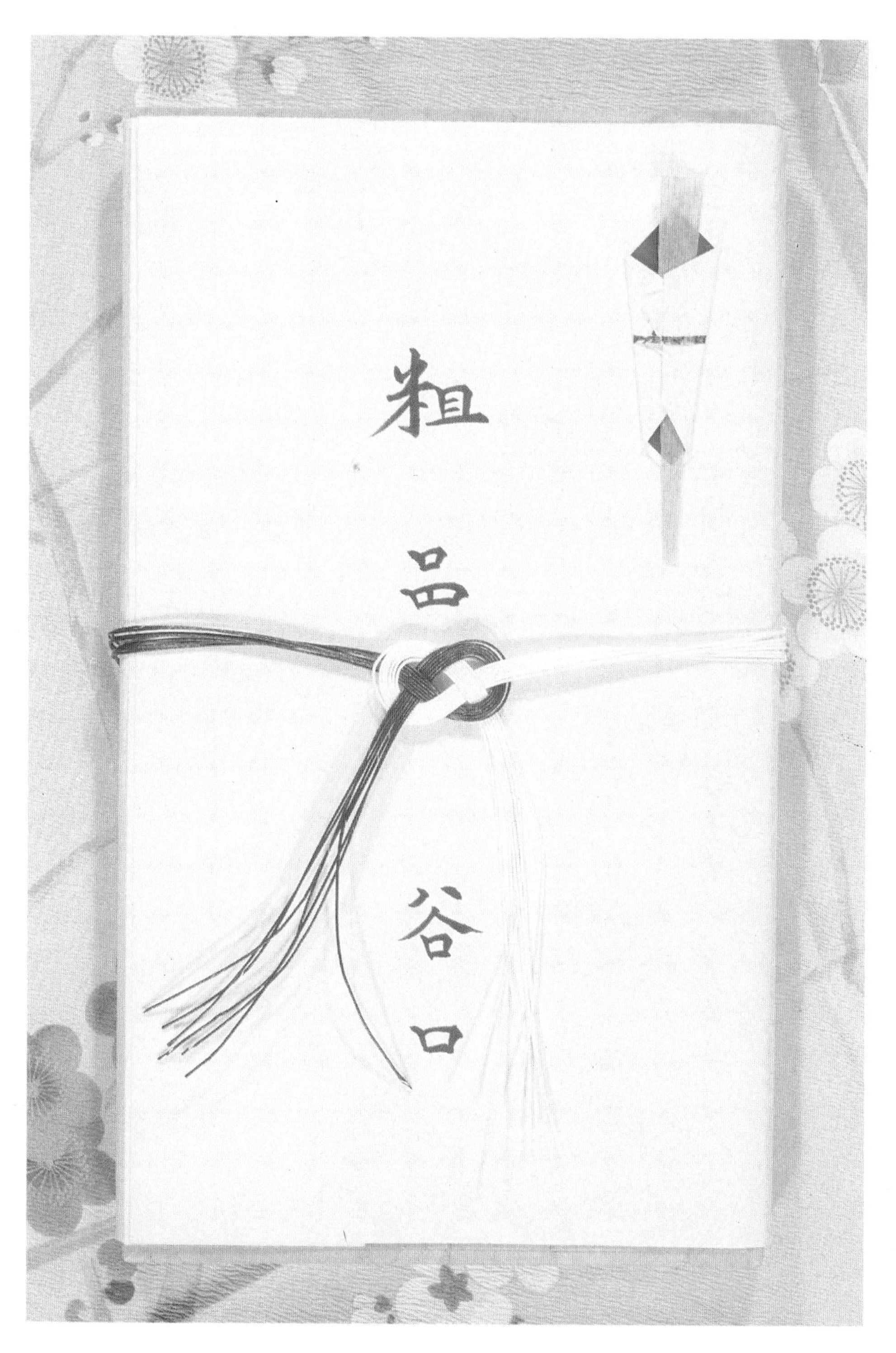

Pl. III. A formally wrapped gift with *noshi* and crimson and white *mizuhiki*; courtesy Oxford Polytechnic. (*See also Fig. 1.5.*)

Pl. IV. Envelopes for wrapping money on happy occasions; the middle one is 'new wave'; courtesy Bob Pomfret. (*See also Fig. 1.7.*)

Pl. V. Three envelopes (bottom) to wrap New Year presents for children, the others (above) to reward participants in a 'full house' theatrical production; courtesy Bob Pomfret.

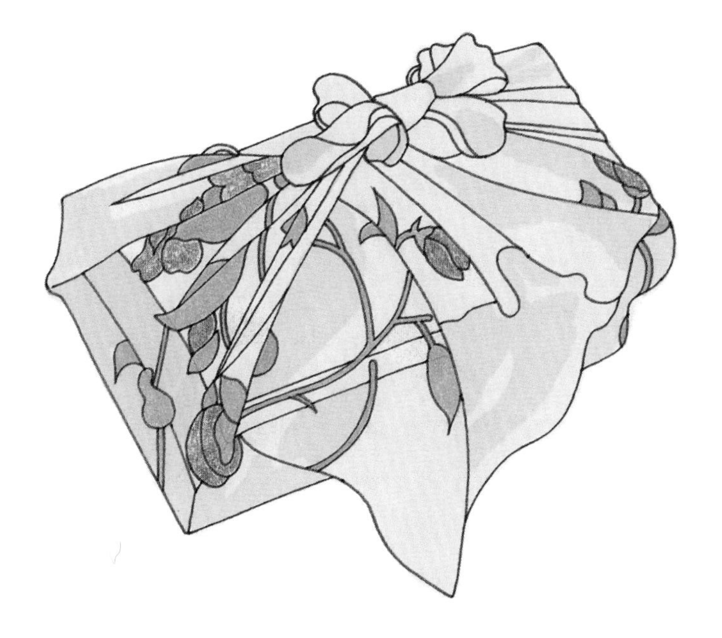

PL. VI. A parcel wrapped in a *furoshiki*; drawing by Lyn North, copyright Japan Library. (*See also Fig. 2.3.*)

PL. VII. Sake served from a red tub into a wooden box ritualizes an occasion; © The Trustees of the National Museums of Scotland 1993. (*See also Fig. 2.12.*)

PL. VIII. A display of packaging for betrothal gifts in a Tokyo department store. (*See also Fig. 3.1.*)

Pl. IX. Kuniyoshi print (*c.*1842) of a bride being dressed; © The Trustees of the National Museums of Scotland 1993. (*See also Fig. 4.1.*)

Pl. X. The twelve-layered *jūnihitoe*, worn by the bride of Prince Aya, younger brother of the Crown Prince; courtesy The Imperial Household Agency, Japan. (*See also Fig. 4.3.*)

Pl. XI. A bridal *uchikake* outer kimono, on display in a Tokyo department store. (*See also Fig. 4.4.*)

Pl. XII. Kunisada print (1858), showing a courtesan with attendant; © The Trustees of the National Museums of Scotland 1993. (*See also Fig. 4.6.*)

PL. XIII. Kunisada theatrical print (*c.*1810–15) showing black-clad stage-hands adjusting the scenery; © The Trustees of the National Museums of Scotland 1993. (*See also Fig. 4.10.*)

Pl. XIV. Kuniyoshi print (1827–30), showing the legendary tattooed hero Rōshi Yensei; © The Trustees of the National Museums of Scotland 1993. (*See also Fig. 4.13.*)

Pl. XV. Kayapo men with Granada Television crew equipment; courtesy Mike Beckham/Disappearing World/Hutchison Library. (*See also Fig. 4.21.*)

PL. XVI. A layering effect is seen in Japanese gardens; this one is at the Hotel New Otani, Tokyo. (*See also Fig. 5.2.*)

Pl. XVII. Kunisada II print (1867) showing layering of space in a courtesan house; © The Trustees of the National Museums of Scotland 1993. (*See also Fig. 5.3.*)

PL. XVIII. Yoshitoshi print (1868); the highest-ranking person in a retinue was carried in a palanquin, often itself wrapped; © The Trustees of the National Museums of Scotland 1993. (*See also Fig. 6.1.*)

PL. XIX. Children often play an important role in festivals. (*See also Fig. 6.4.*)

The use of paper in the partitions of domestic space in Japan would seem to be a very persistent trait, so that even where glass windows have been installed, paper *shoji* may well be found inside them. Tetsuro Yoshida, a Japanese architect, has explained this as being 'to prevent the Japanese room losing its harmonious atmosphere', or, more specifically, 'in order to achieve a subdued light, a warm surface effect and good heat-insulation' (1955: 158). The use of paper for this sort of purpose would seem to date back at least to the twelfth century (ibid. 152), and there is perhaps some significance to Yoshida's contention that the *tokonoma*, which he describes as being at the centre of 'the grouping of rooms' (ibid. 10), has 'throughout the ages . . . been honoured as the sacred place in the house' (ibid. 98; cf. Fujimori, 1990: 15).

The concern with the quality and grain of the wood used in house construction is another aspect of Japanese architecture which may plausibly be compared with notions of wrapping, just as the materials used in wrapping were argued to have value in their own right.[6] Here it is preferred to have unpainted wood, a preference which Yoshida argues could be interpreted as primitive technical ignorance, but he insists instead is 'the natural product of Japanese culture, according to which man should seek to adapt himself to nature rather than subjugate her . . . in fact the result of a highly developed and refined sensibility' (1955: 19). The value placed here on the 'wrapping' of natural wood may perhaps be compared with the value of *washi*, white Japanese paper which, to a Western eye, looks rather crude.

Spatial Wrapping in an Intercultural Perspective

In the Japanese case, there is evidently a parallel between spatial wrapping and other forms already identified, but the language of spatial wrapping may also be examined in an intercultural perspective. Let us start again by looking at domestic space. In the Japanese case I have tried to show that a form of presentation—namely layers of wrapping—which is prevalent in other material and non-material ways in Japanese discourse, could also be identified in the construction of domestic space. In other societies, the construction of domestic space has already been shown by anthropologists to reflect other aspects of their social structure.

Beillevaire (1986) has, for example, shown how the structure of domestic space amongst the people of Taramajima in the Ryūkyū Islands represents in everyday life both the cosmic hierarchy between East and West which governs much of their thought, and the development through time of human relations based on membership in that household. Bourdieu's detailed analysis of the orientation and organization of the Berber house assigns almost every object and activity to a series of oppositions based on the lives of men, outside the house, and those of women, inside, and ultimately concludes that the house in fact represents a reversal, an 'inverted reflection' of the world outside (1973: 110). Again, of the Yekuana people of Venezuela, Guss writes, 'the roundhouse or *atta* is the expression *par excellence* of the Yekuana conceptualisation of the universe' (1989: 163).

A classic analysis of domestic space is Cunningham's study of the Atoni house of Indonesian Timor, in which he argues that the house may be an effective means to communicate ideas between generations, particularly in pre-literate societies (1973: 204). Here, the house is, as in Japan, divided into inner and outer sections, and various rules pertain about who may eat and sleep where, and what these rules imply about social relations, particularly those based on hierarchical differences. Furthermore, the construction and 'order' of the house is analogous to principles found elsewhere in Atoni cosmology and sociopolitical life, although the house is itself incorporated into the wider system. An important distinction between right and left again divides the predominant activities of males and females although an examination of the relationship between sacred and secular space leads Cunningham to identify a pivotal role played by the 'female' category (ibid. 229) in several symbolic ways.

A final example is McDonaugh's (1985; 1987) discussion of the house of the Tharu people of the Dang valley of south-west Nepal. Here, it is argued, the house 'constitutes a kind of template or blueprint for ordering and relating certain ideas and social positions', and 'the language of social structure derives from the house' (1985: 183). As in Japan and Indonesian Timor, the house is divided into inside and outside sections, which are relatively private and public respectively, but another interesting aspect of the Tharu house is the way the most important deities—vital for the definition of the house as a social unit—are housed in the very innermost area, and 'hedged around by rules restricting access to them' (ibid. 184). McDonaugh (1987) provides further details about the rituals associated with these household deities.

A clear distinction between the use of space in British houses and those described for Japan may be observed on a fairly mundane, functional level, in that while there may also be a room—a parlour, or a drawing-room, perhaps—set aside for receiving visitors, the rooms are much more clearly divided according to their purposes than Japanese ones are. Their names reveal these distinctions so that the dining-room is for dining, the sitting-room for sitting, the bedroom for sleeping, and so on. The bedrooms are then allotted to particular people in the house, and it is thought to be important for a child to establish its own identity by having a room where burgeoning personal taste may be freely expressed.

In many Japanese houses, however, the living space tends to be multipurpose (cf. Greenbie, 1988: 16; Yoshida, 1955: 17, 70). The same *tatami* matted area will be used for dining, for sitting, and for sleeping, each at the appropriate time. For the first, a table will be placed in the middle of the floor, usually a low table around which people may sit without the use of chairs, and it will be laid out with the necessary plates and chopsticks. Cushions may be neatly placed out if the occasion is formal, but the *tatami* itself is fairly comfortable for sitting, as in the second case. Some members of the family, usually the head of the household and other men, may have their own chair-backs, with a seat flat on the floor, so that they can lean back, perhaps for watching television. When the time comes for sleeping, all these objects will be stored neatly away and the bedding will be unrolled and laid out in the same area. Here, again, there is some flexibility about who will sleep where, and children are quite likely to crawl in with an adult, rather than retire to their own personal space.

From the outside, too, there are extremely important variations in the wrapping presented to the world at large, and these will indicate subtle differences to members of the same culture which it may be difficult for outsiders to grasp. During the Edo period in Japan, for example, the style of gates and their gables (Fig. 2.14) were regulated according to the status of the owner of the house, with specific rules for very specific differences of rank. These distinctions were based on years of service and levels of taxation (Hirai, 1973: 94). The rules were developed soon after it was decreed that local lords should maintain residences in the capital, for at first these competed with one another to display their wealth, and therefore prestige, in an ostentatious lavishness which got entirely out of hand (ibid.).

Just as the imperial palaces and castles of Japan are surrounded by gardens, moats, and so forth, similar spatial principles have for long separated kings, lords, and squires from the people over whom they exercise power in many other parts of the world, and just as polite language imposes best behaviour, a particularly impressive building will instil awe and careful conduct. There are again subtle differences, however. The front of the home of the Queen of England—Buckingham Palace—is clearly visible from the street, but it is also virtually impenetrable,[7] and its garden is surrounded by very high walls. An attempt to approach it in an unorthodox manner would soon be blocked by a guard. No. 10 Downing Street, the home of the British Prime Minister, was apparently much more accessible, although again manned by a guard, and Mrs Thatcher's installation of iron gates at the end of the road caused some considerable concern about her own evaluation of the role. The northern palace of the King of Norway in Trondheim, on the other hand, gives right on to the main street, and it is said that the King may sometimes be seen riding around on his bicycle. In Madrid, the resplendent 2,800-room palace of the King of Spain is used by him only to receive distinguished foreign visitors; otherwise part of it is open, in a highly regulated manner, to the general public.

At the level of community space, there are again wide variations from one culture to another, and different expectations about its use. Some differences between Japan and Europe have been noted, but there are many other possibilities. Durkheim and Mauss's classic anthropological study about classification (1970) cites Cushing's work on the Zuñi Indians for an example of a people whose method of classifying space is extremely complex and 'remarkable' (ibid. 42), in this context particularly in the way it relates to many other notions which they share. The layout of the village, for example, exactly reflects the social organization of the community, and connections of people, objects, and abstract notions with cardinal points and directions allows all kinds of linguistic allusion clear only to those who understand the system of spatial classification (ibid. 42–54).

In the use of space, too, a community may be relatively open or closed to the exploration of outsiders, and there may also be more or less clear boundaries between the individual dwellings. An Englishman's home is his castle, it is said, and woe betide anyone who strays without any specific business through his front gate. A visit usually requires a knock at the door, or a ringing of the bell before entry, and a prolonged visit may be impossible without a prior appointment. It is unlikely to cause comment if an outsider chose to walk along the street in front, however. In working-class Belfast, on the other hand, it has been reported that it is quite acceptable to walk into a friend's house, without an invitation, or even any specific business, but it is much less acceptable to enter a strange neighbourhood (Milroy, 1987). The thicker layer of wrapping in this latter case is around the neighbourhood, rather than the home. In Japan, the whole country was from 1600 to the middle of the nineteenth century closed off from the outside world so that the nation virtually wrapped itself entirely.

Communities in many of the more temperate regions of the world, particularly those where the clothing worn is scant, often have houses which appear hardly to close up at all. Sleeping arrangements may again be informal, as, for example, in hammocks, and sometimes there are arrangements whereby the men spend the night together in a special 'long house', whiling away the time discussing local politics, perhaps dozing during the day while their womenfolk go about more practical tasks which require the daylight. In some areas, the youths of the community sleep in a special building, or even away in the surrounding bush for a period. These differing uses of space reflect differing notions about the social relationships involved, and it would not seem too far-fetched to seek parallels at other levels, such as language and dress, too.

Finally, in the use of sacred space, there is again a great deal of variety, and a lack of understanding about its nature could inadvertantly cause all kinds of blasphemy. Buildings such as churches, shrines, and synagogues may appear to serve similar roles within their own traditions, but there are subtle and significant differences. Even within the Christian tradition, there is considerable variety in the use of space between branches of the faith, and within the same branch, from one country to another. In some churches, it is only possible to enter certain public areas, and even then various modifications may be expected to one's dress and demeanour. Behaviour will obviously be different, and much more clearly defined during a service, when it may be necessary to know something about the worshipping community before one can even take a seat.

In all these examples, various rules and rituals pertain to the use of space, and this is by no means freely available. Indeed, within a particular society, the ability to use space in the appropriate manner is likely to be one of the ways in which a member of that society expresses their status and, possibly, also their power. It will of course be a variable feature how much any one individual can manipulate the resources available. In some cases, it is necessary to inherit a right to the use of certain spaces—to sit, for example, on the throne of a king—in others it may be a skill which can be achieved through hard work and careful negotiation. In all cases, the use of space is likely to find parallels in the use of language and bodily adornment, as we have seen in the previous chapters. In the following chapter we will examine some of the further social factors involved.

Intercultural Difficulties

In all cases, too, there will be problems for members of a different culture who try to operate in unfamiliar surroundings. As we saw in the Japanese case, there is no small amount of ritual involved in moving through the different layers of a house, around a village, and especially within the compounds of shrines and temples. In some parts of the world where tourists abound, notices will exhort them to observe local conventions in these respects—to cover their arms, legs, or heads, or perhaps to remove their caps, shoes, or sandals—all these to be found associated with sacred buildings within the world of Europe and the Middle East. Further afield, where strangers are rare, their customs may be so different from local expectations that they simply become the objects of ridicule or incredulity, perhaps even regarded by the locals as animals rather than human beings who would surely have some idea how to behave. Or vice versa, as in the case of the many instances of subjugation of a people unschooled in the niceties of 'civilization' as understood by the dominating outsiders.

A visitor may be invited into a house, indeed, in many cultures this would be the expected behaviour, but the visitor may be entirely unschooled in how to behave once inside. It is fairly clear if one must take off one's shoes at the entrance, but it is not at all clear where one should stand or sit once inside. In the Japanese case, there are rules about the use of any room with a *tokonoma*, or special raised dais, because people are supposed to sit in front of it in a certain order (see Chapter 6). There are also non-verbal signals which may be given by the slight misuse of these rules, and even someone who is expected eventually to sit at the top will defer, perhaps for several minutes, before accepting the honour of sitting there.

There are also greetings to negotiate, and on *tatami* matting these may be made from a kneeling position, with a very low bow, which should be kept in position for a period of time depending on the relative status of the participants. At a large gathering, people will get up from their original positions at some point during the event and move round to greet other people present. The use of space is again an important factor here, although the subject of the social relations involved will be discussed in the next chapter. Foreigners, who have no idea of the correct procedure, are usually forgiven for their blunders in Japan, but it is difficult to say how much of this behaviour is consciously carried out, so that a blunder could be taken as an entirely unintended personal affront.

An interesting example of the importance which has been attached to etiquette in Japan is the case around which one of the country's most famous stories is built. The story is that of the 'Forty-seven Rōnin', *Chūshingura* in Japanese, and the incident which left forty-seven retainers without a master, the implication of the English version of the title, was precisely concerned with a matter of etiquette. At the time, the year 1701, ceremonial activity was extremely complex, and very clearly decided. Lords who were expected to take part in such ceremony employed advisers to keep abreast of the rules, but in this case the adviser felt he was being underpaid and he humiliated his master by withholding a vital piece of information. In anger, the lord wounded his adviser with his sword, a misdemeanour for which he subsequently felt obliged to pay by taking his own life—and leaving his loyal followers eventually to follow suit (see Allyn, 1970 for an English version). Such was the power of the knowledge of a detail of etiquette!

To cite a more familiar British or American example, let us consider briefly the use of space at a reception or cocktail party. The chatting which takes place looks fairly casual and informal, and, indeed, it is probably regarded as such by most of the participants. Yet, there are several unwritten rules in operation. First of all, the distance to be left between two people engaged in conversation is a feature which varies from one culture to another, and a person who stands too close by the other's standards may make the second person feel most uncomfortable. Secondly, there are various rules about moving around the room, circulating amongst the guests, especially for the host or hostess. A person who did not understand this principle might feel insulted or abandoned when their allocated time came to an end, whereas another may be regarded as terribly rude for going about the whole thing in an inappropriate manner.

This is a small example, and it could be elaborated in further situations, but in view of the details provided above about the use of Japanese space, it will perhaps already help to make it evident that a fair proportion of the rules of etiquette are concerned with this use of space. Japanese people tend to entertain important foreign visitors in restaurants and bars, where the facilities are less rule-governed, or perhaps more international in nature, allowing them to introduce just enough of their etiquette to fascinate and intrigue, but not enough to allow many layers to be unwrapped. 'Home stays' are a popular way for young visitors, in particular, to experience some of the inner layers of spatial wrapping, as is the au pair system in Europe, but these probably need to be prolonged for the outsider to overcome the layers of politeness which will for a short stay be unlikely to be removed. A superficial impression of Japanese living space has been perpetuated throughout the world as small and poky—the notorious rabbit-hutch image. Small wonder that Japanese businessmen prefer to entertain in the more spacious public areas.

Of particular interest here, as was the case in the previous chapter, is the way one cultural form may impress people from a different culture based on their own ideas of what is impressive. Different people have different ideas about what kind of construction is appropriate, and these ideas may well be related to the climate and resources of their own original habitat, although adaptation to local conditions is of course less important as technology develops. Constructions which show evidence of technological accomplishments, or, better still, perhaps, great accomplishment with apparently little technology, are impressive to most peoples, particularly if they withstand the test of time and become historical or archaeological monuments. Indeed, it would not be going too far to say that buildings provide the impetus to turn a culture into a civilization, recognizable to the rest of the world.

Thus, Japan is evidently civilized in the sense that it can construct beautiful shrines and temples with which to impress foreign tourists, and it even has European-style stations, schools, ministries, and houses of parliament with which to encase the rather different systems which operate beneath them. It can offer palaces and gardens which have inspired horticulturalists the world over, perhaps precisely because of the careful use of perspective, a principle which is widely understood, even if the reasons for it are less immediately clear. Homes, on the other hand, are a layer or two behind the gorgeous wrapping which is usually presented to the outsider, and an entry already represents some intimacy. That this intimacy should precede an understanding of the careful use of space demonstrated by an understanding of Japanese social relations is lamentable, and the comments about rabbit-hutches a supreme example of the need to examine familiar objects with care.

In other parts of the world, the outer wrapping of buildings may be extremely deceptive to those reared in an environment which values first impressions. British travellers, from a country which involves itself in Best Kept Village competitions and requires complicated planning procedures for even the modification of buildings, may find the streets of cities in India and Islamic Africa extremely unimpressive, just as they complain about the gypsies who camp in their vicinity. In all these cases, it is more important to examine interior space to evaluate a family's wealth than to look at the outside of their dwellings. The contrast between outside public squalor and internal private splendour is a characteristic of peoples found as far apart as Bombay and Chicago, but there may well be good reasons for the apparent attempt to deceive, or deflect the interests of the outside world to the riches within.

It is not possible here to examine in any detail a variety of different examples, but it would be an interesting exercise to look at the way clothes and language are used in places which in their buildings have such a deceptive face for the outside world. In Islamic countries, clothes tend to vary in quality, rather than in style, and in many cases they are again effectively covering up most of the person inside, rather than expressing much about their character. The aim here seems to be to conceal, again to deflect the interest of the outside world, although this wrapping will of course have interest in itself. As far as language is concerned, this is an area where sticking to the truth is not only not expected, but even thought to be positively stupid at times. Lies are fabrications, like clothes and buildings, which conceal information reserved for an intimate few.[8] The emphasis is truly on the need to unwrap.

In parts of the world where there is a positive lack of solid buildings, or where the buildings are constructed of local clay—the proverbial African 'mud hut', for example—there is, as was noted in the previous chapter about the lack of clothes, a tendency for those who place great store by such edifices to class the people as primitive and underdeveloped. It is only quite recently that the extremely beneficial and adaptive properties of the mud hut, in keeping the inside cool, as well as allowing ventilation to circulate, have been realized by outsiders so quick to condemn. The reasons for this condemnation are of course clearly related to the almost universal classification of civilization in terms of material objects, especially buildings, which people have acquired the knowledge to construct, and, as we will turn to consider in the next chapters, the centralized power and wealth to employ or force people to build.

Wrapping which Won't Blow Away

This point brings out a special feature associated with the wrapping of space, namely its importance for some less culturally relative reasons. It is about buildings which have outlived their original contents, in some cases for which the functions are only to be speculated or guessed at. It is interesting, for example, that buildings which represent a social order now deeply disapproved of may still be greatly valued in their own right. Thus, ruins of the Roman conquerors who subdued peoples throughout their world, are carefully preserved, sometimes restored, as evidence of our history, whereas the splendour of the palace created by President and Mrs Ceauşescu of Romania served chiefly to shock and dismay.

Stately homes of Britain, whether or not they were built by oppressive magnates who wrung the sweat out of the local people to acquire the riches for their construction, are universally admired for their beauty and elegance. Palaces of Spain unashamedly display the gold they removed from the New World, and the Moorish treasures they acquired during the Inquisition. It could, of course, be argued that many of these 'national' treasures are now open to the inspection of the people at large, and therefore represent to them their own wealth and history, but it is also true that these buildings have effectively outlived their social context.

Throughout Greece, tourists swarm eagerly over the ruins, and in Athens, pillars and other remains are to be found scattered on almost every street-corner. The buildings which represent the culture which was once there are only rarely properly conserved, yet they continue to draw curious foreigners. Indeed, the Parthenon is literally wearing away under their feet. This particular site is the magnet which draws the biggest crowds, despite its dilapidation and, during preservation, the blighting presence of cranes. A short walk away, the site of the market, the Agora, is relatively ignored, despite impressively restored buildings. The Parthenon was the seat of the government, the treasury, the pinnacle of the civilization it housed.

In Mexico and Peru, the ruins of the Mayans and the Incas are the local highspots of ancient culture, although here much less is known about the social life of the inhabitants. We know, however, that their buildings have, to some extent, survived the ravages of time and nature so we admire their abilities. Most of us can make little of the hieroglyphics of their scripts, but we are nevertheless impressed by the fact that they had such a mode of communication. Ultimately, when the people who occupied buildings have gone, we have only the wrapping left to observe. Some scholars do try to unwrap the cultures which they housed, but there is also an almost universal measuring system, which allocates marks of development according to technological achievement and, to a more culturally relative extent, aesthetic skills.

The Ceauşescus were following a well-established tradition by sinking their people's wealth into a magnificent palace. Unfortunately, however, they were too late by the standards of the wider world. Ironically, perhaps, it is the almost naked but feathered Kayapo who have better identified the technological niche of the age in which they find themselves, by becoming conversant with videos and computers. This is also, of course, the area of expertise in which the Japanese have overtaken the rest of the world. In the Japanese case, this was a second chance. They, too, had unsuccessfully tried an outmoded method of impressing the world by developing the top class of their previous social system—the warriors—dressed in the garb of the world they sought to impress.

It took a people who have a great concern with wrapping to work out their mistake here. This chapter has tried to move through the culturally relative to a kind of universal interest in wrapping, expressed in the more or less permanent parts of it which outlive social systems. After all, museums of archaeology may display different objects, but they share certain notions of progress. My argument, however, is that some peoples are so concerned with getting through the layers of wrapping that they may be missing some of the significance of the wrapping itself. At the level of buildings, especially those which have outlived their occupants, we draw close to one another. At the level of language, including the non-verbal, we remain worlds apart. The next chapter steps back towards the realms of language by turning to look at the wrapping of people by people.

6

Social Wrapping, or People Wrapping People

In this chapter we turn to the wrapping of people by people. In most societies, those who command positions of status and power are able, in certain ways, to surround themselves, perhaps protect themselves, by others who will express and confirm their dominant or special role. Even in many egalitarian communities, certain people will play special parts on particular occasions, and these will be expressed by others around them. In complex, hierarchical societies, these forms of expression and recognition can become extremely complicated and elaborate. Patterns emerge and re-emerge in widely separated areas, so this is another topic where we encounter similarities and familiarity in different parts of the world. These patterns may have subtle differences, however, so this area offers yet another opportunity for misinterpretation.

Yet again, too, there are clear parallels in the Japanese case between what I shall term 'social wrapping' and the wrapping principles already identified for other areas of Japanese life. These will be presented first, with examples from several different arenas, both mundane and ritual, and reference will be made to some historical forms of social wrapping which we can all witness nowadays on the celluloid screen. This section will be followed by reference to social wrapping elsewhere, first for the similarities which can be identified, and then, to point out discrepancies and some possible pitfalls, should the signals be misread. In a final section, some more general remarks will be made about the overall implications of considering the arrangements to be described as social wrapping.

Japanese Social Wrapping

Let us start, as in the last chapter, with some illustrations from everyday life. In the offices of many Japanese companies, enterprises, and public services, for example, the desks of almost all the employees are to be found in one large room. Each person may have their own work-space, their own typewriter, even their own telephone, but they will be expected to work amongst the hubbub of all the colleagues seated around them. If there are smaller rooms separated off they may well be for entertaining important visitors, or holding meetings which only a few people are expected to attend. There may also be a few partitions, even individual offices for the section chiefs and managing directors, but these may equally be seated in the same room.

The common principle is the one we can relate to wrapping. The higher ranking people will be sitting at the 'top' of the room, and their subordinates will be arranged, like as not, in the order of their own positions in the company hierarchy. The common entrance to the room is usually at the 'bottom' so that a word with someone near the top must necessarily be preceded by words with a number of people who sit below. The higher up the company ranking system you go, the more layers of people will protect you from the casual, or persistently direct, questions of an outsider who ventures in. In most cases, the outsider will have no need to see the top people, but we will return to this matter later.

This system works well for it has practical as well as symbolic aspects to it. Within a Japanese company, or other enterprise, people usually communicate up and down the hierarchy through those immediately superior or inferior to themselves, indeed special close working relationships are very often developed between specific people linked in this way (Dore, 1971; Nakane, 1973). Thus, day-to-day business is for the most part being conducted by people seated rather close to one another. The chiefs, on the other hand, are seated in places which directly express the higher status of their positions, and serve to remind all those who work with them just how the hierarchy is made up.

This arrangement is particularly important in organizations which insist on the use of *keigo* (honorific language) between the employees, for each person must then use different language depending on whether they are speaking to someone higher up in the room, or lower down. An approach to a senior person must be made through the use of wrapped language, as was explained in Chapter 3, and this wrapping of language nicely parallels the wrapping provided by the seating of the people who use it. It is particularly interesting, then, to find that when the same groups of employees go out to drink in a bar together in the evenings, when their seating arrangements necessarily become more informal, they also drop a number of layers of the wrapped language.[1]

Japanese children learn about these principles of social wrapping at an early age, for a parallel arrangement of employees is to be found in schools. This is also an arrangement with which I have had some personal experience. A typical staff-room will again be organized along hierarchical lines, with a desk for each member of staff, the senior teachers being arranged at the top of the room. In the school attended by my children, the head teacher's room was beyond the top, separated off from the rest of the staff, at the farthest end from the common door. There was another door opening directly into the head teacher's room, but most visitors were expected to approach him through the social wrapping of his subordinates in the staff room. The head's seat was actually positioned at the top of a long table which ran down his room to this other door, so that a person entering for a meeting would again find the head 'wrapped' by those taking seats at the table.

An approach to the school head, then, whether informally or by appointment, was usually to be made through the staff-room. One would arrive at the outer door, where one would be expected to state one's business, even if only to say that the head had summoned one personally. Usually a number of people other than the head would know the business, anyway, and in many cases the matter would be resolved without his intervention, even if it was made in his name. For a straightforward matter, one would need to go no further than the doorway. For a more complicated issue, one may proceed through a layer or two of junior teachers, or assistants of one sort or another, but at some point a person would be reached who is qualified to deal with most matters, and thereby protect the head from disturbance.

The head of this particular school had some experience living abroad, which was not shared by any of his subordinates, and, perhaps as I was the only foreign parent, I did in fact find myself all the way through to his inner sanctum on several occasions. This was particularly useful to me because of my research, as well as the welfare of my children, and I appreciated his time and his concern. I doubt very much whether many of the other parents could have achieved such close contact, however, for there were some 1,300 children in the school and he simply would not have the time. The man who has earned a special office in such an institution has also earned the right to leave most of the mundane matters to those who remain out in the main staff-room. He is universally addressed in honorific language and these wrapped words again express the social wrapping with which he is protected.

In a parallel way, these principles of social wrapping are to be observed amongst the children on some occasions, too. The clearest examples could perhaps be taken from school clubs which typically draw members from all the different years, but organize them hierarchically, according to year. The children make their relative positions within the group quite clear in the language they use to one another and by the terms they use to refer to one another. A senior is a *senpai*, a junior a *kohai*, and behaviour is adjusted accordingly. An example which nicely parallels the seating arrangements discussed above was the way the football club at the school my sons attended took their seats on the coach when going off for away matches. The back seat was strictly reserved for the top class, and the younger ones were expected to sit near to the door. I'm not sure that the others sat in hierarchical order down the bus, though it would hardly be surprising.

We have already discussed the supreme case of the Japanese emperor and the way he and his family are so well wrapped, spatially, that they are almost never seen. In fact, of course, some people must deal with them, and these provide another example of very effective social wrapping. In a study of members of former aristocratic families, who could themselves be seen as a social layer separating ordinary people from the imperial lineages, T. S. Lebra made the point that though their official status was legislated out of existence in 1947, they only really ceased to be aristocratic when they lost their last servant. The 'mediating and, thereby, boundary-maintaining role' of the servants 'kept their status insulated and protected' (1990: 85).

In discussing the lives of the highly secluded children of these aristocratic families, who play within a fenced estate and travel to their élite school in an enclosed carriage, Lebra also notes that the status of the family is often judged by the demeanour and behaviour of the servants who accompany them when they venture out into the mundane world. If the servants are well groomed and properly trained, they transmit an appropriate message to the world, and even impoverished aristocrats would spend money on the servants' clothes while maintaining a frugal family table (ibid. 85–6). These children's servants were also responsible for training their charges in manners and etiquette so it was important that they should set a good example.

A personal and intimate account of the imperial households in immediate post-war Japan is provided in English in a book by the American tutor to the Crown Prince at the time, now the Emperor Akihito, aptly entitled *Windows for the Crown Prince* (Vining, 1952). The veritable battery of chamberlains, ladies-in-waiting, imperial guards, and members of the near nobility make it virtually impossible under normal circumstances for those outside the system to communicate with those inside at all, and one of Vining's explicit duties was to provide her charge with some windows to the outside world.

The extent to which the Japanese imperial family are so well wrapped that they are virtual prisoners within their own palaces was brought home to those of us able to meet Prince Naruhito, now the Crown Prince, at the time of his study visit to Oxford. During his time in England, he was much freer than normal to walk about in the outside world, accompanied only by a single policeman, and he expressed very poignantly his joy at being able to do this. His rather solitary, self-conscious figure could be seen on many an afternoon threading his way through the Oxford streets, his bodyguard not far behind him. A formal approach needed still to be made through the social packaging, however, as I learned to my cost.

During this time I organized a meeting to bring together anthropologists interested in the study of Japan, a group which subsequently became known as the Japan Anthropology Workshop. On the first evening a Japanese entrepreneur in Manchester had agreed to bring a Japanese meal all the way down to Oxford to welcome our guests, the cost arranged, by him, to be covered by a Japanese company operating in the north of England. It was never made explicit why this man should be so philanthropic, but I suspect it may have had something to do with our intention to invite the visiting prince to join us on this occasion, a sure way, he probably thought, to bring publicity to his enterprises.

As planned, I did invite the prince, who had informally expressed interest in the organization and the event, but I did it in entirely the wrong way, it would seem. He was a member of Merton College at the time, and I simply sent him an invitation via his pigeonhole. For a long time there was no reply, but eventually a letter came, typed rather inexpertly, and signed 'Naruhito' by himself. He was sorry, but he would be unable to attend after all. Afterwards I learned that all the prince's public engagements were arranged by his chamberlain. Evidently I had misunderstood the social wrapping involved, and the Japan Anthropology Workshop had to put up with a very disgruntled entrepreneur. We do own a letter signed by the man who will probably become the emperor, however!

This is just one example of how social wrapping protects those it envelops, and how important it is to work through the appropriate channels. People in certain positions must be approached through the people who surround them, usually the people down below. If they were too accessible, too unwrapped, their position would be weakened and the whole system would eventually be destroyed. To some extent in Britain the Japanese prince was being allowed to lead the life of an ordinary student, and fellow students were able to make arrangements with him rather informally. I suspect that our problem lay in the fact that we were too close to Japanese life. Our meeting was, indeed, reported in the European newspaper for Japanese people living here. For reasons such as these we should have operated in a Japanese way.

An interesting parallel at the linguistic level here is to be found in the language used by Japanese people living in Oxford who were presented to the prince during his visit. Many of these were people who would not normally have had an opportunity to meet members of the imperial family and they had little idea how to address him. There was some discussion about this before the occasion, and the chamberlain was able to offer advice, but several of the Japanese people resorted to English when they found themselves face to face with this young prince. I had great sympathy with these people as I had done the same myself. I knew very little of the language suitable for addressing members of the imperial family, but in any case his face and demeanour were those of a youth, someone of an age with one's students (in my case), and I found it impossible to slot into a suitable Japanese mode of address. It was an easy way out to use the language in which he was being allowed to be more free.

This apparent lack of freedom raises questions about the actual power of the Japanese imperial family, indeed, of the emperor himself. This has been an issue of some international interest, brought to a head with the death of Emperor Hirohito in 1989, but the views of Japanese people on the subject are very relevant in the context of this book. The person of the Japanese emperor is not that important to them, they say, it is the symbolic role that he plays that is significant because he stands for the Japanese nation and the Japaneseness of the people who live there.[2] Until he was persuaded by the Allied Occupation to renounce his divinity, he could perhaps be seen as a supreme point of contact with the spiritual world, the ultimate *yorishiro*, but like the *shintai* in the centre of the shrine, which represents the deity, the chief role of the emperor is the representation of the people, their ancestral roots and their history. This is a point to which we will return.

In the past, when the castles of Japan were occupied by their local lords, there was another form of social wrapping which reflects the spatial wrapping discussed in the last chapter. This was the way in which the people who lived around the castle were arranged by proximity according to their rank. Aristocratic estates would be in the immediate vicinity of the castle grounds, along with shrines and temples, and the houses of lesser retainers would be a little further away. Generally, the higher the rank of the warrior, the nearer to the castle he would be allowed to live, and sometimes an extra moat would surround the houses of the very highest ranking ones (Hirai, 1973: 14, cf. Alex, 1963: 26; Motoo, 1986: 28).

Within these castles, status differences were again clearly expressed through the seating arrangements, the highest-ranking samurai sitting nearest to the lord, the lowest nearest to the entrance again (M. Inoue, 1985: 109). The *tatami* floors were of different levels in accordance with these rules, and even the thickness, shape, and texture of the cushions would vary as one drew

FIG. 6.1. Kuniyoshi print (*c.*1844) showing troops marching into battle (of Nagaragawa *c.*1350)—a visual expression of the power structure; © The Trustees of the National Museums of Scotland 1993. See also Pl. XVIII.

further away from the master's seat. He himself would often sit on a couple of extra *tatami* mats so that no one even approximated his status (Hirai, 1973: 70–3), and he would be completely enclosed and protected from the outside world by rows of retainers, their levels growing gradually lower as they receded from him.[3]

When eminent people such as these lords moved around in Japan they were again surrounded by huge retinues of subordinates to protect them from the ordinary populace, who were anyway expected to bow most humbly as they passed. These retinues would form public expressions of the power structure (Fig. 6.1), again a form of wrapping of the people in positions of status or power, which separated them off from the ordinary people and gave them an aura of unapproachability. Very often the lords, themselves, would be carried in a palanquin so that they would not even be visible (Plate XVIII and Fig. 6.2). For battles, of course, the lower ranking warriors would protect their leaders from the skirmishes in this way.

In the modern Japanese world, parades may still reproduce these historical retinues in the form of festivals, which allow people to witness and participate in activities resembling those of their forebears. In the provincial seaside city where I lived during my most recent period of fieldwork a local festival had been created only a few years previously which required adults and children to dress up as samurai warriors (Fig. 6.3) and parade up a hill to the site of the erstwhile castle. In fact a reproduction castle has been built there as a tourist

FIG. 6.2. Palanquins were used to carry members of samurai families.

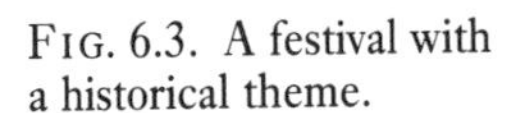

FIG. 6.3. A festival with a historical theme.

attraction, and the purpose of the festival, organized by a group of local business leaders, was also largely to attract and amuse visitors. It did this by portraying scenes from local historical events, or, more accurately, from a well-known story (and TV film) based on local historical events.

The events portrayed recount some of the political history of the area, further elaborated in the local museum. To use a festival for this purpose seemed to me a little strange at first, because festivals are more often associated with Shinto shrines or occasionally Buddhist temples, but in fact they do also provide public expressions of local allegiances. In virtually every community in Japan, there are annual festivals for local shrines, and the usual form is for local bodies—the youth group, the children—to pull or carry a portable shrine around the neighbourhood (Fig. 6.4), collecting money as they go. This will eventually arrive at a particular spot, sometimes together with portable shrines from other regions of the area, and a ceremony will be held which involves important local dignitaries.

The vehicles themselves carry no local lord, although they may carry a few children playing flutes and drums, but the whole event is organized from above by local men of power. These men draw up plans for the route of the shrine, and they will await its return. They also give sums of money which may be written out and attached to the shrine for all to see. As in the previous chapter, the portable shrine represents, rather than carries the local deity, and it is to be admired from without, rather than penetrated to some sacred core. Moreover, the very fact that this shrine is taken out to the community perhaps emphasizes the importance of the social and spatial wrapping which the deity commands.

FIG. 6.4. Anja Festival, Tokyo; courtesy Japan National Tourist Office. See also Pl. XIX.

For modern ritual occasions, seating arrangements are also fixed quite clearly so that those who are playing the most important role are seated in front of *tokonoma*, with others in order down from there, as mentioned in the previous chapter. Thus, for example at a wedding, the bride and groom will take pride of place, along with the go-betweens, and the guests will be seated in order of closeness and hierarchical relation to the family, age, and so on (see Hendry, 1981*a* for further detail). At a funeral, the chief mourners will sit at the top, again representing relations between their own houses and those of the deceased (ibid.). If no one is playing a particular part of this sort, the usual convention is to sit in age order, generally with men above women.

An important point which can be raised here is the relativity of the seating arrangements depending on the occasion. Within a particular social community, whether it be residential or based on common employment, the same people may sit together for different events in quite different hierarchical relationships to one another. The boss will not always sit at the top because he is the boss, although that is his position in the office, and all other things being equal, he will usually be placed pretty high up. An important outside guest would take precedence, for example, and so would one of his employees who was getting married.

The seating represents roles being played rather than universal differences of status, and 'reading' the seating arrangements, or the social wrapping, provides important information about the event taking place. In fact, sometimes the principals may be entirely absent, as in the case of the presentation of betrothal gifts, mentioned in Chapter 3, and at funerals in Japan, it is not necessary for the coffin to be present, either. There will usually be a photograph of the deceased in pride of place, and it is in front of this photograph that the farewell burning of incense takes place. Again the centre of the proceedings is a representation of something rather than necessarily being anything of any very great significance in itself.

Social Wrapping Elsewhere

There are of course versions of social wrapping found elsewhere which may be directly compared with these Japanese examples, and some of them are rather similar, at least on first sight. The organization of office space may vary somewhat from the Japanese prototype described above, but it is certainly true that in most parts of the world one must approach the senior members of an organization through their subordinates. It is, for example, a recognized measure of advancement to acquire a secretary to answer one's telephone, and those whose rooms open out both into the corridor and into the secretary's room may well insist that visitors be screened by taking the second of these two routes.

A visitor may assess his or her own status depending on how far the socially wrapped person they are visiting comes out of the inner sanctum on their departure. As Raymond Firth has pointed out, to be seen off at the desk, the door of the inner sanctum, the door of the outer office, the door of the lift, or even the outside door of the building, are 'recognised status indicators'. A secretary or assistant may also be dispatched to see a visitor out. Firth makes the point that 'the degree of disarrangement of the one party indicates the relative status of the other' (1972: 20).

In large and complex organizations, in Japan as elsewhere, there will be a telephone exchange and a whole reception staff to deal with outside enquirers and direct them to the appropriate department, although these same enquirers may resort to all kinds of subterfuge if they feel they are being fobbed off with the wrong level of response. Within any particular social system, people acquire an understanding of how their version of social wrapping works, and they will become more or less skilled at negotiating, or unwrapping it. If the reason is important or pressing enough, people in our Western system may feel that they will be happy with no one but 'the top'. There is an assumption that if one can only get to the top, problems will somehow be solved, and those who reach the top positions are expected to be able to respond accordingly.

Elsewhere, this may not be the case. The structure of companies in Japan is in fact disarmingly familiar, for many of the job titles are translated into terms which are quite recognizable. However, their meaning in practice may be somewhat different.[4] Japanese industrialization was of course modelled on industrialization in the West, but much that is more indigenously Japanese persists below this Western wrapping, and the modes of unwrapping are also distinct. Indeed, the whole assumption about having to reach the top—the essence or the centre—is again misplaced here. There is a person at the top, to be sure, but the role he (or she) plays is not the same. As in other Japanese arenas, this person is representing something, in this case the company, and there will be roles to play in this capacity. This role may not include solving the same problems as it might elsewhere, and the qualities required of a person who reaches the top may also be different.

An important point to be made in this context is that the wrapping a particular organization chooses to present to the outside world may, of course, be as much of a front as the polite language used by its representatives in their dealings with the public at large. It follows, then, that a similarity of wrapping does not necessarily imply a similarity of internal order, and this is certainly the case in comparing Japanese organizations with the Western ones which may, at some point, have served as models. The formal power structure is one thing, determining a degree of status for those who command positions in it. The internal power structure may be something quite different. Much ink has been spilled on the subject of Japanese management,[5] and I have no intention of competing here with this abundance of material. The point is simply to emphasize the way principles of social wrapping may parallel those found in other arenas, both in Japan and elsewhere.

To turn again to schools, in the West, too, there is a hierarchy of teachers, and the heads usually have their own rooms. Here the hierarchy is different, however, and it is not usually necessary to negotiate the staff-room to reach the head's inner sanctum. Indeed, one of the duties of school heads is to liaise with parents and potential parents so these are the people visitors are actually most likely to meet. I had an appointment with the head of a school in Scotland once, and, on arrival, I asked a pupil the way. There was a system of reception, phoning up, and so on, but I had entered by the wrong door, and the pupil simply took me straight to his room. In fact, I was sent back down again, briefly to observe the appropriate protocol, but the lack of reserve on the part of the pupil I found quite interesting.

In Europe, too, there are royal and noble families who are sometimes still surrounded by retinues of servants and attendants, although increasingly few in most countries. Here, as in Japan, the incumbents often hold very little actual power, although their status may be high. In most cases, too, their countries are ruled by democratically elected governments, and the seats of political power are to be found within these institutions. The royal families play roles rather comparable to those of the imperial family in Japan, in that they represent their people in a ceremonial and symbolic way. The value of the role may perhaps be witnessed in the fact that so many such families continue to prosper, and to maintain the social wrapping which makes possible their continued special role.

Just as the presentation of their palaces varies, however, the nature of their social wrapping will also almost certainly be different. It is quite possible, for example, for ordinary people to catch glimpses of the British Queen and members of her family if they attend the appropriate social events, and garden parties are held annually at Buckingham Palace to which people who have made an impact in various walks of life are invited. Indeed, members of the British royal family make a point of speaking to people who turn out to functions they attend, and they will shake hands with all sorts of onlookers as well as those they have been scheduled to meet. Many of them are said to be very friendly and approachable on these occasions, indeed, it is part of their training to know how to speak to anyone and put them at their ease.

The affection of the British public for particular individuals bears little relation to the precise role they play, as has for several years been witnessed annually on the Queen Mother's birthday. Members of the same family who are less careful quickly acquire a reputation, perhaps for being snooty, or flighty, and this is then widely discussed in the press. In Britain, much more is made of the individual incumbents of the various roles than there is in Japan. The 'royals' become personalities whom the local populace like to emulate.

They frequently feature in the media, and they are expected to appear regularly in the world at large.

In the world at large, however, they may find themselves subjected to different forms of treatment depending on the varying views of their roles taken by peoples of different parts of the Commonwealth. In Polynesia, for example, they may find themselves being carried from one place to another so that their feet do not touch the ground. Local chiefs are given such treatment, and this is indeed necessary, for if they were to walk on the ground they would immediately make it taboo for all commoners to use (Steiner, 1967: 40).

Over the years the social wrapping which accompanied the various European explorers and empire-builders in fact probably played a significant part in impressing the peoples whom they were able to subdue and control, especially where there were similar forms of wrapping already in existence. The ability to impress through forms of dress has already been discussed in Chapter 4, but the support of retinues of attendants, servants, and subordinates of one sort or another undoubtedly contributed to the peaceful domination of local peoples, where this took place.

In the New Guinea Highlands, for example, Gordon and Meggitt identified a pattern of 'ceremonialism' as being partly responsible for the effectiveness of the government of the Australian patrol officers (1985: 177–81). In 1970s Southern Ethiopia, on the other hand, a Japanese anthropologist greatly impressed a couple of his British counterparts, managing themselves in a modern way by adapting to local custom, when he and his exquisitely dressed wife passed by with a long train of bearers whom they had engaged to transport their luggage and house-building equipment into the field (André Singer, personal communication).

Again, some of the impressions which were created may well have been through trappings of status and power which have superficial similarities in different parts of the world. Sitting on raised mats at the raised end of a series of increasingly elevated parlours is a fairly direct representation of superior status, as is the raised floor which accommodates the 'high table' in European colleges. Thrones are fairly clear indications of special positions, as are the palanquins and carriages which transported people in certain positions in the days before limousines and private planes. Parades and retinues of one sort or another make direct statements to the attendant public about whatever they happen to represent.

These are important symbols of status and power, designed in part to maintain that status and power, wherever they are found. They may again carry subtle differences, however, and there is one well-known Japanese example which illustrates the persistent theme of deflecting interest away from the centre. It is the story depicted in the film entitled *Kagemusha*. This recounts the tale of a Japanese lord who was injured and subsequently died at a crucial moment in a series of battles in which he and his followers were engaged. His immediate subordinates, sure that a knowledge of this information would undermine the confidence of the army he headed, giving the opposition a tremendous psychological advantage, decided to pretend that he was still alive.

They were able to do this because the lord had a double, a criminal who was spared punishment on the condition that he comply with the deception. The double sat in the lord's palanquin and on his raised mats for two years before he was eventually found to lack a distinguishing birthmark by the lord's former nanny. Much of the story centres around the person of the criminal, but the point I want to emphasize here is that it was actually thought possible for a person completely without experience or power of any kind to sit for two years in the seat of the lord and leader at the pinnacle of this particular social group. Important decisions were evidently made by people elsewhere in the hierarchy and the role of the person of the lord (in this case the criminal) was again to represent himself, and therefore the group, without necessarily having a great deal of practical power or influence. In this case, the battles proceeded rather well until the unwrapping went too far.

There is quite clearly a need for someone to be in the top position. The other famous story of the forty-seven masterless samurai, *Chūshingura*, discussed briefly in Chapter 5, confirms this. The retainers and followers of the master experienced many problems in trying to organize themselves once the master was dead, and only a part of the original number was strong and loyal enough to carry out vengeance on their master's behalf, before following him in ritual suicide.

In the case of festivals and more secular parades, we might turn briefly to consider how one or two European examples compare with the Japanese *matsuri*, a term with connotations of worship, and older ones of government, although usually translated as 'festival'. A direct comparison with the historic festival would perhaps be provided by the historic regatta held annually in the city of Venice. This is a parade of boats, representing particular periods and events in Venetian history, and it is held during the height of the tourist season. In 1989 there was a political point to be made, however, which is rather interesting in this context.

During this parade, there is usually a point at which a fleet of gondolas carries the current city dignitaries past the waiting crowds, a few moments of glory for the incumbents of these elected offices. In the autumn of 1989, however, the gondolas were present but they carried no one inside them. They were paraded empty down the Grand Canal. Now, it is not necessary to know much about the local political situation to recognize this as a form of protest. These city dignitaries usually passed in front of the waiting crowds, wrapped in the historical splendour of their city. On this day, the gondoliers left without them, and they were, in their absence, shamed before their people.

In a way, this protest has a nice Japanese touch to it. The wrapping goes on, with or without the current incumbents of the positions of power. The city and its history provide a symbolic basis for Venetian consciousness, shared by the people who elect the local dignitaries, as well as those persons themselves. If the practical politics of a particular period fail to satisfy the people for whom it is supposed to benefit, they can speak through their shared wrapping about the impugned status they feel the so-called dignitaries should be assigned.

In most other European countries, together with American and other previously colonial nations they influenced, there are parades from time to time, which may well offer similar possibilities for the people to express themselves. They generally feature bands and marching, and a series of floats which provide an ideal venue for such communication. The Lord Mayor's show is a British example, with no religious content, which takes place in cities such as London and Oxford. Again, it is not really the Lord Mayor who is important, but the city which is depicted in the various floats. Those who decorate and ride in them have an opportunity to display the various organizations they represent in the way they feel appropriate. It is a day when they can exercise some pretty evident influence in the definition of their city.

New Views of Wrapping

In fact, in recent years there has been a curious reversal of interpretation of social wrapping in the so-called democratic world. Unlike the anthropologists, trying gamely to adapt to their African surroundings, members of erstwhile imperially expansionist nations now seek in different ways to impose their new systems on the rest of the world. Condemning all other solutions, including and perhaps especially those which resemble their own previous regimes, they proselytize through economic and sporting sanctions, except in extreme cases where they may still resort to more violent means.

The egalitarian ideology which underpins this system has been so enthusiastically absorbed (at least at a theoretical level) that it invokes a kind of spontaneous disapproval of anything hierarchical, a disapproval which also prevents an easy understanding of systems based on long held notions of difference. This point, made some time ago by Dumont in his classic exposition of the Indian caste system (1980), is just as valid as the one made in the previous two chapters about people who lack recognizable forms of material wrapping. Our capacity for misunderstanding is great, and always conditioned by views currently fashionable about the way things should be. However, perhaps taking steps towards recognizing this bias may help to overcome it.

7

Temporal Wrapping—and Unwrapping

The final example of forms of wrapping is concerned with the use of time. This is not a discussion about conceptions of time, though it is conceivable that it could be. It is about how time is divided up. We will be concerned with the way in which events are separated off from other events, and with the structuring, in time, of those events themselves. We will also be concerned with the way in which temporal ordering may allow a degree of unwrapping, providing a way through layers of formality to less formal layers beneath, and the possible benefits of building up the layers again.

This chapter will bring together ideas formulated in previous chapters, demonstrating parallels between different types and levels of wrapping, and it will lay the final foundations for the following chapter which will examine in more detail the relationship between power and wrapping, and the role the use of wrapping may play in the influence and manipulation of others. Like previous chapters, this one will start by examining Japanese examples of temporal wrapping and then go on to look at other possibilities.

Beginnings and Endings

A fairly clear and straightforward example of temporal wrapping in Japan is to be found in the way events are separated off from the time surrounding them by quite marked beginnings and endings. This is true of a wide variety of events, from mundane, everyday occurrences to the grandest ceremony. The form of these beginnings and endings is clearly decided by society, so that they could qualify to be classed as ritual, although sometimes it is difficult to distinguish them from what one might better describe as 'routine'. Some of them become part of a longer sequence of activity, which can be compared with the various layers of wrapping, so perhaps it is better to forgo further classification and concentrate on the notion of wrapping again.

An example which illustrates the beginnings and endings rather well is a practice which occurs without fail at the start and finish of classes of one sort or another in Japan. This includes classes for children and adults alike, in music, art, sport, and a variety of other accomplishments. The teacher and pupil or pupils will first of all assemble, perhaps in a fairly informal manner, until all are ready to proceed. There may be some conversation, exchange of news and ideas, and even questions and answers directly related to the activity in question. Before the class will start properly, however, there must be a greeting, which is followed by a formal request by the pupil(s) for the teacher to proceed. An expression of appreciation will terminate the proceedings at the end.

The importance of this little routine/ritual is made particularly clear when it is being demonstrated to children. A good example is the first lesson of any particular accomplishment, for most of the time may well be devoted to practising and perfecting the opening sequence. My younger son had been learning the piano before we went to Japan for an extended period of fieldwork, and I decided to hire an organ and engage a teacher with the hope of continuing his progress. In fact I was quite disappointed, for the teacher seemed entirely unable to build upon his previous accomplishments, and insisted that he start again and follow her usual pattern—perhaps an example of wrapping in itself—but the point of mentioning this here is to describe the first lesson.

The organ was ignored entirely for most of the fifteen-minute slot, and quite a substantial number of minutes were spent by the teacher trying to get William to sit up straight, bow to her, and pronounce the formal request, *onegai shimasu*, which is the way to begin any lesson. A Japanese child of 6 years would have done this without hesitation, I feel sure, for they would already have acquired the two pieces of cultural knowledge which lay behind it, namely that any such activity requires a formal beginning, and that these are the words which are expected. The teacher could simply have worked on the style and manner of delivery. Unfortunately William was entirely unprepared, and even refused to do it, so that the teacher was reduced to continuing that part of the lesson for several weeks until he gave in, probably out of sheer boredom with the idea!

In fact ritual phrases of one sort or another form an important component of an adult's concerns when she is engaged in the business of socializing her small children (see Hendry, 1986: 73–5), and several of these are precisely about dividing up time and separating off events from one another. A very mundane example is to be found at the beginning and end of every meal, when the first mouthful should always be preceded by an expression of receipt, namely *itadakimasu*, and the end of the meal with a phrase alluding literally to the feast it has been, *gochisōsama deshita*, but serving more as an expression of thanks.

Children are usually so used to this routine by the time they enter kindergarten or day-nursery that they will sit for several minutes, without touching a morsel of the food laid out in front of them, until everyone is ready and the phrase has been pronounced. Nor will they get up and leave the table until the end has been formally enunciated. In kindergartens there is actually often a more elaborate beginning to meals and break-times, when children on duty will run through a longer sequence of questions, with answers from the class, about whether everybody is sitting properly, has been served, and is ready to begin. The words are entirely fixed, although the sentiments are part of the training to care for others around, as well as one's own needs.

Kindergarten is a place where beginnings and endings become more clear and elaborate generally. Although children will arrive at a variety of different times, there is invariably an opening sequence before the events of the day begin, and several ritual ways in which the day is divided up. At one kindergarten I visited during research on early child rearing, the headmaster sat outside the school formally greeting each child as they entered the playground. More often there is one clear routine for each class, and this will very likely include the teacher playing a fixed tune on the piano to announce that children must assemble in an orderly way, possibly the singing of a morning song, and ultimately the reading of the register.

A specific example involved the following sequence of activity. The teacher sat down at the piano and played the usual morning tune, a signal to the children that they must sit down in a circle in the classroom. Those who were still outside would come in at this point, removing their outdoor shoes on the way and donning indoor ones, as well as changing into their daytime smocks. Once everyone was seated, the teacher would play a drill which would signal that the children should rise to their feet. This was followed by a cadence, to which each child would make a deep bow, and then all would break into a song of morning greeting. A spoken greeting would complete the fixed routine before the teacher checked up on each child's appearance, and eventually read the register.

A similar routine precedes meals and break-times, and the children go through another sequence of parting rites at the end of the formal part of the day. It is not unusual to devote about half an hour to preparations for going home. At one Christian kindergarten I visited, the children changed their clothes and then spent a substantial period singing and praying in unison, engaging at appropriate moments in fixed movements of the hands and body including yawning, feigning sleep, clapping, waving, and shaking the teacher's hand, all interspersed with the stacking of chairs, and the distribution of books and papers to be carried home.

Adults do not abandon all this training in their daily lives, although they naturally truncate some of the children's elaborate routines. Thus, most Japanese will utter the word *itadakimasu* before embarking on a meal, even in a Western setting, and they are usually rather punctilious about greetings. An interesting parallel to the way children arrive at kindergarten and then only later carry out the formal greeting may be seen when visitors arrive at a house for a specific purpose. They may enter the house, using the series of small rituals mentioned in Chapter 5, sit down, and even perhaps receive a cup of tea. When the member of the household they have come to visit appears, however, their business may well be preceded by a further formal greeting.

I once arrived at a house (on time) before the man I had come to visit was properly dressed. He carried out much of the rest of his toilet in front of me, since his mirror was in a part of the house where I was seated, also spending several minutes greeting the godshelf and making the morning offerings which undoubtedly marked the start of his day. At last, when he had completed all this 'beginning', he came to sit down with me and we bowed formally to each other and went through the usual exchange of greeting pleasantries as a 'beginning' to the discussion which followed.

All manner of adult and children's gatherings are marked with formal beginnings and endings, some of which will be described in the next section, when further elements of temporal wrapping are examined, but it might be interesting to point out another aspect of this propensity formally to mark off periods of time in the way the Japanese year is divided up. Japan has imposed both the Chinese and the Gregorian calendar on its own earlier system, and the result is a veritable series of layers of openings and closings.

The most marked occasion today is the New Year which coincides with that of the Western world, when several days of holiday follow a mammoth attempt to terminate all outstanding business, pay all bills, make presentations to benefactors, and celebrate with workmates a formal closing of the year. The New Year is marked with special food and drink, dressing up, presentations of gifts and money within the family, and visits to shrines to seek an auspicious start. Acquaintance is formally renewed through a special greeting, delivered in person or posted on a card, and several further gatherings bring together relatives, friends, and workmates as they celebrate the opening of the year.

Although the beast associated with the Chinese year is officially ushered in on the first of January in Japan, there is, around the time of the old Chinese New Year, another small ritual which is generally associated with a kind of purification rite. This is *setsubun*, when beans are thrown out of the house to cries of 'out with the demons, in with good fortune', and housewives and farming folk begin to talk of the spring. This day is in fact the official beginning of spring, known as *risshun*, but it is in early February and it is still some weeks away from warmth and the opening of buds.

There are several other rites associated with spring, and each represents another kind of beginning. The academic year starts on 1 April, as does the financial year, and it is usually just before this time that cherry blossoms appear and parties of associates take food and drink out to the parks and sit under their ephemeral beauty. It is now that new clothes, desks, and stationery must be acquired for children starting out on new phases of their academic careers, and it is at this time that the annual strikes are held to make clear to companies that their workers are not to be taken for granted.

In the countryside, there is a marking of the beginning and end of the agricultural cycle when the rice god is said to come down from and eventually return to the mountains. This is a way in which the cycle of farming activities is 'wrapped' temporally, although there are several other rites during the specific seasons of crop cultivation, notably when rice is transplanted and harvested. The details of these beginning and ending rites vary somewhat from region to region, but they exist throughout Japan. The French anthropologist Laurence Caillet describes them as 'a reflection of the ecological complementarity of spring and autumn, and of the symbolic complementarity of the human and divine worlds' (1986: 42).

Though there may well be more examples of this propensity to mark beginnings and endings throughout the Japanese year, let me refer to one last example, as a clear imposition of cultural categories on a natural cycle of seasonal changes. This is the beginning and end of the 'bathing season' in the seaside town where I carried out two extended periods of field research. Although the weather may become quite hot and sticky enough seriously to contemplate sea bathing before 1 July, this is the date officially deemed the beginning of the season, just before which the beaches will be cleaned up, and no one ventures into the sea before this time. The end of the season, at the end of August, when all facilities are rigorously removed, is apparently accompanied by the arrival of jellyfish which would anyway make further swimming inadvisable.

It could be argued, then, that the Japanese year may be seen as a series of ritual openings and closings. Indeed, the words used in Japanese for various beginnings often imply an opening. As the year proceeds, layers of activities are unfolded, so that this offers a parallel with layers of wrapping in other ways. Approximately in the middle, there is the Bon midsummer festival, when ancestors are said to return to visit their descendants, an occasion clearly complementary to the New Year celebrations. At this time, too, gifts are sent round which are described by a word combining the character for 'middle' with that for 'beginning', also used for the first day of the year (*chūgen*).

Unwrapping the Layers of Temporal Wrapping in Japan

Lest it appear that the argument about wrapping is becoming a little forced at this stage, let me turn now to support the idea of temporal wrapping with some examples of the 'layering' of time, and an examination of the way in which 'unwrapping' may be associated with layers of decreasing formality along a temporal axis. The examples given here present a framework, to be developed in the next chapter, about the power wrapping may confer, and advantages of the ability, alluded to already in the previous chapter, to operate in a formal context.

To start with a fairly straightforward example, when people meet strangers for the first time, they are usually constrained to behave in a formal manner, especially if these are people they are likely to encounter again. The exchange of name-cards between Japanese businessmen is well known. These provide all the necessary information about the holder's workplace, and his position in a hierarchical schema which is universally understood, so that this allows a precise level of formality to be determined. Otherwise people need to be cautious at first, usually choosing over-formal terms of address until they have ascertained some information about each other. They are then able gradually to adjust to a mutually acceptable form of discourse.

As people get to know one another better, however, they will very likely be able to drop some of the formalities (cf. Minami, 1987*b*: 25), especially in informal surroundings, such as the bar after work. Those who have known each other for years, perhaps since they were at school or university together, will feel at greater ease to speak freely and express their views frankly. Indeed, many Japanese people meet their old childhood associates from time to time throughout their lives, and find the occasion a good one for unburdening themselves of secret feelings they would refrain from revealing to even their closest daily contacts. Interestingly, gifts presented to the most intimate of one's contacts are very often wrapped in only a minimal way.

Another expression of this intimacy is also to be found in the way groups of colleagues, co-students, schoolfriends, and so on will go on trips together. This very often involves spending a night or two in a resort, perhaps by the sea or at a hot spring, when much of the activity is carried out in the informal *yukata* and slippers provided by the hotel. Members of the same party will bathe together, in a large communal bathing area, although usually segregated by gender, and they will also very often sleep in rows in a large room rather than splitting up into individual cubicles. This is reminiscent of the way people bathe and change into *yukata* when they return from work to the intimacy of their homes. Some families are also inclined to sleep in one large area, rather than individual rooms.

I carried out some research once with Japanese housewives living in Oxford, an area of particular interest for me being the language these women used with each other. They were a very good example of the point being made here, for most of them came with their husbands and met each other only for the first time in Oxford. They were all quite clear about the way they modified their language gradually as they became better acquainted with one other, reverting to more polite and formal tones if they felt relations were proceeding too fast. Those who became most intimate with each other were also using the most informal language, although not necessarily the least polite, since this was associated with other factors, such as upbringing and previous social experience.

An interesting extension of this idea, incidentally, may explain the phenomenon several foreigners visiting Japan have noticed, namely that Japanese friends sometimes reveal the most intimate secrets about themselves when they are speaking English. This foreign tongue is evidently associated with an informal level of communication, perhaps influenced by the idea that Westerners, typically Americans, are supposed to be frank with each other, and maybe even the apparent informality of their clothes, from a Japanese point of view. It is, of course, also well known in Japan that foreigners, on the whole, are pretty casual about the wrapping and packaging of gifts!

This notion of 'unwrapping' associated with time is an oversimplified model, which is so far little different from the situation found anywhere as people grow to know one another better. The ideas will be developed further in the next chapter, however, where unwrapping will be examined as a political process. Here let us first consider some further examples of the model.

Japanese meetings can provide another case in point, although of a slightly different order. This is also a case where misunderstanding is very likely at an intercultural level. The idea, prevalent in the West for some time, that groups of Japanese people make decisions 'by consensus' is patently open to suspicion, and I would suggest that this is simply a misreading of a layer or two of wrapping. It is quite possible that meetings may appear to achieve unanimous decisions—indeed, they very often do—but this must simply mean that these decisions have been reached somehow before the meeting ever started. Even a cursory investigation will uncover the practice known as *nemawashii*, the literal 'tending of the roots', or groundwork, which precedes most meetings in Japan, as indeed elsewhere.[1]

In practice, within a meeting, there may be a point at which some decision-making takes place, but it is important for participants to identify when this is. They must wait until a suitable number of layers of formality have been unwrapped. A PTA meeting I attended will provide an example. It opened, as might now be expected, with formal words of greeting (*aisatsu*). These made use of fixed polite expressions, with very little content, about the weather and the busy lives of those who had made the effort to be present. They bore little relation to the subject-matter of the meeting, but they opened the occasion to proceed further.

The next part of the proceedings continued in a slightly less formal tone, now running through the actual items to be discussed at that day's gathering (*hōkoku*). There was still little opportunity for discussion, influence, or decision-making, but the subject-matter had been broached. The event had been defined. It was only in the third part of the meeting, known as *giji*, that discussion became part of the order of the day. Interestingly, at this point, the language used became much less formal, and local dialect was heard much more than at the earlier stages. Here was the chance to make some comment, here people could be heard from the floor, and there was time for an airing of views before the 'wrapping up' reformalized things again.

An even clearer example of this type of temporal unwrapping, as we should perhaps call it, may be discerned in Brian Moeran's (1984) analysis of sake-drinking sessions in a pottery community where he worked in Kyushu.[2] Moeran identified five distinct stages in the proceedings of these occasions, each one allowing progressively more informal interaction, and he argues that full participation in these events is a vital part of political life in the community concerned. Disagreeing with another widely held idea that Japanese people forgive and forget home truths and criticisms which may emerge under the influence of alcohol, he suggests that though such indiscretions may not be mentioned in more formal sober life, they are positively stored up for any future advantage which may occur.

The first of these five stages corresponds with the beginnings mentioned earlier. People actually gather gradually before this stage, chatting in an informal (inside) part of the house, perhaps sipping tea and eating snacks. When all are assembled, they move through to the formal room(s) prepared for them, where they take seats according to gender and relative age, ranked in order down from the *tokonoma*. This is the moment for the *aisatsu*, the short, fixed greetings, thanking those present for taking the trouble to attend, and, in this case, announcing formally the specific occasion in hand. The seating arrangements express a formal hierarchy of households, which Moeran later distinguishes from the locus of actual political power.

The following four stages take the participants through increasingly informal periods, when there is gradually more likelihood that people will speak 'from the heart' (*honne*), rather than as a matter of public 'principle' (*tatemae*). During the second stage, seating positions are held, but seating posture is relaxed, and there is a chance for people to exchange cups, and conversation, with those around them. During the third stage, people leave their seats and move around the room to engage in the same activity, which Moeran sees as quite directed and often politically significant. The fourth stage interrupts the flow of conversation with bouts of singing, and the fifth becomes quite drunken and rowdy, sometimes belligerent.

Moeran notes that at each stage the type of discourse changes, allowing greater freedom to move towards matters unmentionable during the normal course of everyday life, matters which may not be mentioned afterwards, but which he argues will surely be remembered. As in the PTA meeting mentioned above, the language moves from the standard speech of the formal greetings through to increasingly informal use of local dialect. The use of space will also reflect the increasing informality, and the final stage may even move back into the family space from which the gathering originally emerged. The posture of the participants also becomes more and more relaxed so that some will even fall into a drunken sleep at the side of the gathering.

That meetings should follow a fixed order of this sort is another matter which is learnt as early as schooldays in Japan. Within school life there are many occasions for ordering events in an expected manner, and they are organized and carried out largely by the children themselves.[3] One occasion I attended was particularly revealing because it was a meeting during which a further event was being planned. The group holding the meeting consisted of all the children in the school from a particular neighbourhood, a type of group which meets termly to agree on rules they should observe *en route* to school as well as within the neighbourhood, and discuss events they may hold during the holidays.

Because the group comprises children from each class in the school, with elected members of the top class in overall charge, the children learn the proper procedure from each other as they move up through the six years they are there. A teacher is usually present at the meetings, but she takes an active part only when members of the group become disruptive. The meetings again start with a formal opening, and there is a standard pattern they follow. At this stage there would seem definitely to be a point at which ideas may be raised from the floor, discussed, and, if approved, these are written down on a board and later typed up for distribution.

The Christmas party being planned at the meeting I attended was also expected to follow a particular pattern, and this was mapped out on the board before suggestions were put forward for the optional parts. There was little or no discussion about the pattern, for this was agreed implicitly, with a formal beginning and end, and there was even a rather limited selection of possibilities for the concrete activities. Nevertheless several ideas were mooted, and discussed, and decisions were eventually made and entered on the board. I had the distinct impression that the older children wielded much more influence in the decision-making, and indeed, some of the younger ones grew bored by the whole event. This drawback is overcome on other occasions by dividing an event into elements taken care of by each year in the school.

The Wrapping and Unwrapping of Japanese Ritual

In Chapter 5 various rituals were described for entering and penetrating the inner depths of a Japanese house, for 'unwrapping' layers of domestic space. An idea was also given of the complexity of penetrating the grounds of the imperial palace. Similarly, approaching, or 'unwrapping' the increasingly sacred centre of a Shinto shrine or Buddhist temple is accompanied by a series of rites: rinsing the hands and mouth (Fig. 7.1), wafting incense over oneself (Fig. 7.2), casting a coin into the box, and removing shoes. In Chapter 5, however, the point was made that rites may well be completed at a suitable point in the wrapping, that reaching the centre is hardly ever the object of the exercise, and unwrapping usually only proceeds so far.

An alternative explanation of the apparently wrapped approach to Shinto shrines and Buddhist temples is that the path all the wrapping marks out is an architectural representation of the 'path' or 'way' religious practitioners are

FIG. 7.1. A rite of purification while approaching a shrine or temple; courtesy Japan National Tourist Office.

FIG. 7.2. Wafting incense also takes place outside the temple; courtesy Japan National Tourist Office.

supposed to follow. The word Shinto is, of course, literally 'the way of the deities', and Buddhism can also be described as a 'way' to enlightenment. Bloch discusses this idea, pointing out that the most sacred building then 'proves on inspection to be merely a continuation of the alley way', eventually 'a narrow tunnel to the beyond'. He also notes the way the path equally provides an entry, at least in the case of Shinto deities, for the spiritual power to come to the inhabited area (1992: 51–4).

There are various ways to access this power, however. Those who clap their hands and mutter a prayer at the entrance to the shrine area are making a direct bid for attention. As described in Chapter 3, they may speak from the heart in a way that requires no special wrapping, and the offerings they make are also unwrapped. Even a monetary gift is simply cast into a large receptacle, which stands in front of the shrine, quite in contrast to the way money is wrapped up to be presented in secular encounters. Some will write their requests on a little wooden plaque known as an *ema*, but even these are hung up outside for all the world to read.

For a special occasion, a priest may be asked to mediate in attracting spiritual assistance, and then a ceremony may be held inside the building itself. Priests also approach most closely the sacred objects at the innermost sanctuary of these religious buildings and the deities they represent. However, they must be prepared in quite elaborate ways. They may have been born to the task in the first place, as heir to the priestly role, and they will also have

FIG. 7.3. Several *torii* close together give the impression of a tunnel.

spent time studying, and preparing in ritual ways, for the more esoteric tasks of the religious specialist. Interestingly, however, this preparation brings them into more formal, rather than more intimate, relationships with the deities. Buddhist priests may shave their heads, but they and their Shinto counterparts are expected to wear elaborate garments, and to chant entirely fixed prayers and supplications so removed from the speech of everyday life that they are incomprehensible, if recognizable, sounds to the lay observer.

This activity can hardly be called unwrapping, then, and it contrasts with the approaches of ordinary people. I consulted some high-ranking priests about whether the fixed language of *norito*, or Shinto prayer, could be compared with *keigo*, perhaps thereby expressing formality in front of the gods, but they denied any similarity, arguing that the language of prayer was fixed because it was ancient, dating back some 1,200 years. This language is powerful, however, as was discussed in Chapter 3. The words may not themselves be comprehensible, but their value is as a form of wrapping again, and this time the power of the deity is actually invoked by means of wrapping.

In fact another way to access spiritual power is actually to close oneself away, as was briefly described in Chapter 1. The ascetic practice of seclusion known as *komori* involves a sort of literal self-wrapping in a closed room, a temple, or the darkness of a cave, usually accompanied by fasting.[4] This practice was interpreted by the folklorist Origuchi (1945) as drawing on a notion of power associated with a sealed vessel, the idea of a sacred force gestating and growing because it is shut away. This notion has been analysed at some length by the Dutch anthropologist Ouwehand, who points out that the concept of *utsubo* basically means empty, but at the same time it implies, 'the empty in which, invisibly and supernaturally, a divine principle resides or can reside' (1964: 123).

This is reminiscent of Grapard's description of medieval pilgrims who travelled away to sacred mountains, for he describes them as 'walkers in emptiness'. When they left home, he explained, they moved into a realm which transcended their former knowledge of the world. Their 'walk' paralleled the 'path' of Buddhist practice. Such a practitioner did not reside anywhere, he writes, he was 'one who abides in emptiness' (1982: 206). This idea Grapard relates to the teaching of Esoteric Buddhism that the original residence of the Buddha is within one's own heart and mind, but this must be freed from illusions and passions (ibid. 208–9). Again, it must be emptied.

The teachings of Buddhism are beyond the scope of this book, but I find interesting the value evidently placed here on the notion of emptiness. Throughout the preceding chapters we have been directing attention away from unwrapping, back into the surrounding packaging, and now we have reached religious ideas which seem to urge us in the same pursuit. One last quotation from Grapard's study is somewhat impenetrable, as it stands, but it certainly seems to support our efforts to break away from isolating an essence within any particular package. He is writing of the Esoteric Buddhist practice of 'entering' a mandala: 'The practitioner goes from the manifestation to the source, from the form to the essence, and finally reaches the realization that form and essence are two-but-not-two' (ibid. 209).

The value of emptiness identified here can perhaps now be invoked to explain, at least in part, some of the findings of previous chapters: the gifts which appeared to be little more than wrapping, the folded paper as a protective talisman, powerful words without any literal meaning, white garments as a *tabula rasa*, uncluttered space at the heart of the house, and the depiction of Tokyo by Barthes and Bognar as the city with an 'empty centre'. Of course, the emperor abides in this empty centre, but if we hypothetically return to him his divinity, this would fit the pattern well. The next chapter returns to the question of any practical power the emperor may be said to possess.

There is one final example of temporal wrapping and unwrapping in a ritual context which it will be useful to bring up at this point, and this is the sequence of events which takes place during a tea ceremony. There have been various attempts to explain the tea ceremony as religious activity (see e.g. Anderson, 1987), and there are undoubtedly elements in it from the various religious paths prevalent in Japan. Ultimately, views tend to depend on the definition of religion which is adopted, and this is not really an issue here. There can be little problem in my view, however, with describing the ceremony as ritual, and some of the analysis is certainly parallel to that observed in a religious context.

The tea ceremony is manifestation of a practice or 'way' comparable with other Japanese arts, such as archery, kendō, and flower-arranging, passed from generation to generation through the diligent imitation of the actions of a leader or teacher, to whom a pupil becomes attached. Each element of the practice is clearly fixed, although there are slight variations between different schools, and there are more and less elaborate versions of the ceremony within each school. In each case, there is an order of proceedings which is followed rigorously (Fig. 7.4), and there is only very limited opportunity for choice of detail within this structure, itself heavily influenced by the season and more specific time of the year.[5] In the context of a class, the variety will usually be selected by the teacher, otherwise by the host/ess.

A foreigner coming to the tea ceremony for the first time may find the whole thing extremely tedious, possibly quite painful, for it is essential to sit on one's feet in a formal position (Fig. 7.5) throughout the chief part of the procedure, which, even in its shortest form, lasts long enough to inflict discomfort on those unused to the kneeling position. Most of the action is carried out, slowly and deliberately, by the person serving the tea (the host/ess), who gives the guests a chance to move only twice: when they are presented with their cake, and when they receive their bowl of tea. Even then,

FIG. 7.4. Each movement for making the tea at a tea ceremony is clearly fixed; courtesy Japan National Tourist Office.

FIG. 7.5. It is necessary to sit still for long periods during the tea ceremony; courtesy Japan National Tourist Office.

the movements are clearly fixed. To consume the tea, for example, involves taking and turning the bowl so that it faces out, a clear manner of sipping, an examination and admiration of the pottery, and a precise returning of the bowl to the host/ess.

All this is only a small part of the complete cycle of events shared by real initiates in the 'way of tea', however, and an examination of a more complete process introduces some clear parallels with the rites accompanying an approach to a Shinto shrine. Participants might well, for example, go through a series of rites of purification, in some cases completely changing their clothes, often rinsing their hands and mouths at an appropriate water source provided in the garden (Fig. 7.6), and always removing their shoes in order to leave the immediate pollution of the mundane world outside the entrance to the tea-room.

An example of a rather elaborate, but standard version of the ceremony, the noon *chaji*, a three- to five-hour event towards the hosting of which training is designed to proceed, has been described and analysed by two anthropologists (Kondo, 1985; Anderson, 1987). There are several differences of inter-

FIG. 7.6. Rites of purification may precede more elaborate versions of the tea ceremony; courtesy Japan National Tourist Office.

pretation, or, to Anderson, fact, in their descriptions, but both illustrate very clearly the way a precise order of activities moves the participants from their mundane everyday lives into and out again of a series of ritual activities, during which they are offered the opportunity of sharing a rather profound spiritual or transcendental experience.

Apart from sharing several types of beverage, served in a variety of ways, and consuming a meal together, the participants are invited also to enjoy the beauty and tranquillity of a freshly watered garden, to examine and admire single examples of several forms of art, and, for several hours, only to engage in minimal verbal conversation. Put like this, the whole event could still sound rather boring, but most people who have taken up the practice of tea, a practice requiring innumerable hours of just such activity, describe instead the calm, the transformation, perhaps even the therapeutic effects they eventually begin to experience.

Kondo describes the various activities and movements involved as a symbolic journey from mundane space and time, through stages of greater and lesser formality, to a ritual climax which offers the participants the possibility of a 'distilled form of experience set apart from the mundane world' (1985: 302). She examines in turn the role of auditory symbols, as opposed to verbal modes of communication, and the role of the various substances involved, in her efforts to 'deconstruct the ritual'. Anderson (1987) objects to Kondo's use of the term 'formality', but she is convinced that the ceremony is a religious activity with soteriological goals which she proceeds to analyse in some detail. She shows how it responds to goals not only of Zen Buddhism, with which it is often associated, but also with those of Taoism, Shinto, and Confucianism.

Kondo uses a metaphor of 'unfolding' to summarize the events of the ritual, and it is difficult to know whether to accommodate these descriptions to a temporal axis proceeding with unwrapping (of the mundane worries of everyday life) or wrapping (in the apparently extreme formality of the clothes, verbal and non-verbal language, space, and time). Furthermore, interestingly enough, in neither description of this particular version of events, is there a particularly clear rite of beginning or ending, comparable with those we have so far outlined. Rather a sequence of activity emerges which does certainly move in and out of ritual and mundane modes.

Folding is, of course, an important part of wrapping. The metaphor was also used by Bognar in describing the 'folded' space of Japanese cities (1985: 67). Since the tea ceremony brings together many Japanese arts, as well as religious symbolism from a variety of sources, all of which impinge on the five human senses, I feel it is important to consider it most seriously. In fact it points to the complexity of the subject. Until this chapter we have kept considerations of wrapping rather straightforward by looking at each manifestation in relative isolation of the others, but gradually we have come to build up parallels which run across the board. Taken together, these examples begin to demonstrate the tremendous potential power which this wrapping bestows.

Actually, the tea ceremony is an excellent example of many things. The casual foreign perception is, as usual, way off course, partly because it is usually based on a very limited experience, possibly also influenced by local (Western) notions of tea parties. A deep knowledge of the practice is not easily obtained. It takes years of regular participation, and even after that, it would seem that foreigners can reach no ready agreement about its role, as Kondo and Anderson demonstrate. An ability to carry out a proper tea ceremony, in all its elements, requires a great deal of knowledge of a variety of artistic pursuits. The power and influence such knowledge imparts is also beyond easy estimation, but the high regard in which this accomplishment continues to be held, as well as its continuing popularity in many areas of Japanese life, hint at considerable significance, spiritual and otherwise. In the final chapter we will try to put this question of power, and its relationship to wrapping, in an intercultural context.

8

Politeness, Packaging, and Power

This chapter will bring together aspects of previous chapters which have been concerned with the relationship between wrapping and power. The use here of the term 'power' is very broad. It is concerned with the control people have over their own lives and those of others around them. It is concerned with the extent to which an individual, or a group he belongs to or represents, is able to impress and influence another individual or group, and the possibilities one person or a group may have of manipulating the world around them. It is concerned, too, with the way 'the world' may take advantage of specific individuals and groups in the pursuit of apparently corporate goals.

This chapter will attempt to put the Japanese case into an intercultural context. We will return first to Japanese examples of linguistic wrapping and, drawing on some of the anthropological theory on language and politics, we will look at how this compares with reports elsewhere. We will then move on to emphasize the importance and political significance of looking at other forms of wrapping, again drawing on comparable cases from cultures other than Japan, but eventually suggesting ways in which we might learn from the exercise carried out in this book.

The Power of Linguistic Wrapping

In Chapter 3, where Japanese forms of politeness, known as *keigo*, were described as linguistic wrapping, several allusions were made to the power a skilful use of such language may bestow. At the most basic level of interpersonal relations, for example, it was noted that an appropriate use of *keigo* could help to persuade people to do something they may not otherwise have planned to do by making them feel good about it. The example was cited of the Kyoto man who became Japan's number one confectioner, apparently at least partly by using polite language in a meaningful way. Using politeness as a form of strategy is nothing new. Indeed, it is one of the underlying assumptions of the most detailed and scholarly attempt to date to seek universals in the use of politeness formulas (Brown and Levinson, 1976).

At an even more basic level, the power associated with simple forms of greeting, is already well documented. In the previous chapter, greetings were identified as important aspects of creating significant beginnings and endings for events, leading into a form of unwrapping which would allow the drawing out of opinions, the possibility for participants at a gathering to influence the proceedings. These greetings were polite, and they were also highly ritualized in that there was very little choice about how they were to be made.

A paper by Raymond Firth includes a strong argument for considering 'greeting and parting behaviour' in general as a type of ritual because of its patterned routines: 'it is a system of signs that convey other than overt messages; it is sanctioned by strong expressions of moral approval; and it has adaptive value in facilitating social relations' (1972: 29–30). Such assertions may be tested quite easily by anyone who has the courage, or temerity, to try interfering with socially sanctioned greetings, and then gauge the effect this interference has on their associates. Just such an experiment was carried out by the linguist Charles Ferguson, who simply failed to reply to the 'good morning' greetings of his secretary for a few days. The 'terrible effect' he reports on the office indicates something of the 'power' of a greeting (1976: 140).

Goody identifies three functions of greetings in her study of the Gonja people. They are, first, 'to open a sequence of communicative acts', secondly, 'a means of defining and affirming both identity and rank', and, thirdly, 'a mode of entering upon or manipulating a relationship in order to achieve a specific result' (1972: 40). The first, probably a universal function of greeting, corresponds with Japanese 'beginning', the second, found wherever the form of greeting varies with status, also encompasses the Japanese case in the formal distinction of status. The third leads us beyond these examples, however. In the Gonja case, this is a type of greeting which precedes a request of some sort, or what Goody refers to as a 'greeting to beg'.

A further, clear, and detailed example of this last function of greetings, is Irvine's (1974) analysis of Wolof greetings as strategies of status manipulation. She makes the important point that the person who initiates a greeting, very often a long series of fixed exchanges, is putting themselves into an inferior position to that of the person addressed. If two people catch sight of each other, they are also under an obligation to greet, so adjustments of this sort are being made continually in Wolof society. The advantage, however, of getting one's greeting in first is that a superior is obliged to respond positively to requests of various sorts from an inferior. There may thus be attempts during the course of the exchange to reverse the order suggested at the outset.

This example is, in fact, also an illustration of another form of temporal wrapping, as well as of the power of greeting rituals, and Irvine's Wolof material offers a variety of interesting points of comparison with the Japanese case. The underlying atmosphere of competition implied by the greeting possibilities is, for example, reminiscent of another aspect of *keigo* mentioned in Chapter 3, namely that it may be used as part of a battle 'between people who use language in order to gain an edge over those around them in constant struggles for ascendance in the informal hierarchies of status and prestige'. *Keigo* was associated in that chapter with the idea of self-defence. It was suggested that it is used as a kind of protective armour, but also as a weapon.

In Chapter 3 the argument was in fact based largely on a discussion about the language of women, which is interesting in view of the fact that the language of Japanese women has, like that of women elsewhere, been described as an expression of their inferiority and their lack of power. Feminist writers argue that women are socialized into being tentative, polite, and feminine in their language use, all qualities which deny them the possibility of being assertive and powerful in social relations (e.g. Lakoff, 1975). Akiko Jugaku, who has made a detailed historical study of Japanese language and women in a book of that title, confirms that Japanese women are expected to learn feminine linguistic skills in their efforts to be feminine, or *onnarashii* (1979: 5–62).

The correct use of *keigo* is certainly related to this quality, but my research would suggest quite the reverse of the feminist argument. In relations between women, it was clear that facility with the use of *keigo* conferred considerable power on the user to impress and influence others (Hendry, 1985, 1990*c*, 1992). The research of linguists has also shown that the language of women varies very much with the context in which they are using it so that the appearance of difference between men's and women's language stems as much as anything from the different frequencies of interaction of men and women in different situations (Ide, 1986; Shibamoto, 1987). Thus, the arguments put forward about women's language could also plausibly be applied to the language of men.

Another general theory about the differing language of men and women is that women are more polite because very often they are in less secure social positions than men, and their language is therefore more important for the establishment and maintenance of their status. While men are allocated status according to their occupation or earning power, women are more likely to be assessed according to their appearance and general demeanour, including language (e.g. Trudgill, 1972: 182–3).

In Japan, women as well as men in companies, universities, and other clearly defined occupational hierarchies are able to use at least some of their *keigo* in ways dictated by these positions, but women in housewifely encounters are much more likely to be constantly assessing and even manipulating each other by their use of *keigo*, on the basis of which they also allocate status and prestige to one another. It is my contention, however, that these less formal[1] uses of *keigo* are in fact found throughout society for struggles in the informal hierarchies which operate alongside the formal ones of occupational relations. Thus, the actual use of *keigo* in everyday life reflects not only these formal differences of status, but informal personal differentiation and manipulation, used by men as well as women, but about which one can learn a lot by examining the language used by housewives.

The example of drinking sessions, discussed in the last chapter as a form of temporal wrapping and unwrapping, illustrates this distinction between formal and informal use of power and status in a Japanese political arena. Moeran's interpretation of events is that the 'official', 'formal' authority of the elders, expressed in the seating arrangements in the early part of the proceedings, is gradually being undermined by the practical, informal power of the middle-aged men who actually have much more say in 'vital community matters'. I suggest that we may in fact be looking at two different levels of wrapping here. That of the elders is more associated with status, to be sure, and that of the younger men, with the wielding of practical power, but this seems rather appropriate in a Japanese situation, as has been pointed out in various discussions about the role of the emperor.

As Moeran himself points out, much of his discussion ties in with the work done by anthropologists on political oratory in traditional societies. He compares the difference between the public speech of *tatemae* and the more private *honne* which is revealed (or unwrapped) during drinking sessions with the two types of 'tight' and 'loose' Maori traditional oratory (1984: 240), described by Anne Salmond (1975), and the 'crooked' and 'straight' speech of the Ilongot, described by Michelle Rosaldo (1973). I think the comparison can usefully be taken further.

Salmond's (1975) work does indeed divide types of Maori traditional oratory into two, one more formal than the other, and she emphasizes that the first is concerned with prestige and the second with the power to influence more practical matters. She also describes a third type of political arena, with different skills, namely that of the New Zealand government. The first of these three is highly structured, with a strong element of ceremonial. Participation is limited to those, mainly elders, who have already acquired a fair degree of prestige. The last, which is very influential in the daily lives of Maori people, is open to much younger Maori, through their education and European experience, but these same nationally political figures tend to be deferential and respectful at the traditional gatherings. The 'loose', middle type of political oratory, open to any who care to participate, provides a training ground for either of the others, and a point at which local politics may be entered and influenced.

Clearly we are dealing here with different types of power, in this case achieved and expressed through different types of oratorical skill. Each has an appropriate venue for its exercise, and those who wish to be involved need different qualities and skills to excel. The three are by no means mutually exclusive, but they represent different arenas of Maori affairs. As Salmond notes, if only one of these arenas was considered, it would give a very partial view of things. Her example is that if one looked only at the 'tight' structure, it would seem to indicate an authoritarian political structure governed by an elderly élite. In European countries, we are accustomed to considering the two more practical arenas as the seats of political power, relegating monarchs and the like to an apparently relatively powerless ceremonial role. However, I would like to push a little further here with the role of this last type of what we might call 'ritual power'.

At the end of Chapter 3 we considered the special role played by people able to use certain restricted forms of language, comparing the use of Latin in European churches and the chanting of Shinto and Buddhist priests with South American ceremonial dialogue, and noting a certain similarity with the way skilful use of *keigo* may be seen to define people as members of a type of élite. Since *keigo* is also the formal language used to effect the various openings and closings of meetings and ceremonial occasions, the prestigious role of carrying out these ritual parts of events is limited to those conversant with its use.

If we concentrate our attentions only on the parts of these gatherings when people appear immediately to be influencing the proceedings, i.e. the type of political activity we are used to in Western systems, I think we may well also be missing something important. This would perhaps be equivalent to our overriding interest in unwrapping things, in getting to the essence of something. Just as Salmond warned us not to stop at the ceremonial role of the elders, we must also, on the other hand, be careful not to underestimate the power of the ritual elements of proceedings. Indeed, we must try not to throw away any layers of wrapping before we understand more completely the role they may be playing.

In the previous chapter the elements of temporal wrapping were described for various Japanese gatherings, and it was suggested that following the order of them allowed a degree of unwrapping to take place. In other words, discussion of issues needed to be preceded by a certain amount of ritual procedure without which the discussion would probably not take place. The Samoan *fono* public assembly would seem to follow broadly parallel lines, starting out and ending with the sharing of *kava*, moving in between into a series of ritual speeches, with little opportunity for discussion, and then proceeding to a period when language may be much less formal and opinions may be expressed (Duranti, 1984). Duranti identifies and discusses two different types of oratory, which he describes, following Firth (1975), as in the first case 'mostly homiletic, confirming the already known', and in the second 'persuasive and manipulative' (1984: 231).

In fact, Firth's (1975) study of the Tikopian variety of *fono* is interesting in this context because he points out that although decisions may be made about calling a *fono* by the chiefs, they are not necessarily the ones who will make speeches. Indeed, much of the work of summoning people and setting up the *fono* is carried out by their executives, the *maru*, to whom the role of speech making is also delegated. Although the chiefs wield the power in Tikopian society, according to Firth, they are not even supposed to attend the assembly because they are too sacred. The *maru* are their mouthpieces, instructed to speak for the chiefs, which, in case of dissent, saves their status from being impugned in their presence (ibid. 35).

Evidently, it is necessary to look further than the speeches themselves to discern the locus of power in this situation. A parallel example for Japan was described in Chapter 6 when it was pointed out that local festivals are carried out by many people, dressed up in a variety of ways, but those who have made the important decisions about the event, and given most money, are not to be seen in these public displays. Instead, they are waiting in relatively sombre garb inside a shrine building, or another suitable location, for the retinue to return. In Japanese, there is a concept to describe the exercise of power from behind the scenes, which also uses a theatrical metaphor, namely that of a 'black curtain' (*kuromaku*), sometimes also translated as 'wire-puller'. It is used particularly in reference to the rather sinister activities of far right wing characters whose worlds dovetail closely with that of organized crime (e.g. Kaplan and Dubro, 1986).

Bloch's overall thesis in the introduction to *Political Language and Oratory in Traditional Societies* (1975) has been criticized in several places (e.g. Borgström, 1982; Brenneis and Myers, 1984; Irvine, 1979), but he does make some interesting points relevant to this discussion. In the first instance, for example, he points out that the influence from other social sciences on political anthropology has not always been beneficial because it leads anthropologists to focus on data comparable to that obtained in these other fields so that they may lose the special value of anthropological speculation—which might arise as a result of the confrontation of foreign cultures. This sounds like quite a good explanation for the bias of interests I am trying to identify and overcome here.

Bloch goes on to talk about the relatively hidden political behaviour he was able to observe, particularly as an outsider, which was permeated throughout the social intercourse of the members of the society in Madagascar, where he worked, and which was determined by the hierarchical relationships accepted by the people as an inherent part of their social order. This is the traditional authority which makes up the burden of social control, found in all societies, 'permeated through the constant requirements of such things as respect to the elders, and . . . appropriate polite behaviour . . . Most of the time . . . this appears to the actors as the only natural way to behave' (1975: 4).

The socialization of children in appropriate polite behaviour struck Bloch as interesting in Madagascar because of the way the focus of correction and direction of behaviour was not so much on the content of what was said but on the manner in which it was said. He compares this with the attention paid in English society to the use of words such as 'please' and 'thank you', as well as suitable intonations of the voice which are thought of as respectful and not 'cheeky' (ibid. 5). Bloch saw the reason for this attention to regulating the manner of speech as concerned with a concomitant (hidden) restriction, in a subtle and effective way, on the content of what was said.

The formal behaviour which Merina children gradually learn in this way equips at least some of them as they grow up to participate in the formal exchanges which Bloch identifies as important elements of social control in Merina society. Bloch argues that the formality, by its very nature, 'dramatically restricts' what can be said, and, more importantly, perhaps, the responses to it. Once a formal situation has been created, he argues, it is almost impossible to object to proposals which come out of it. The ability to create such formality is evidently a *sine qua non* of being able to manipulate it, and only those who become skilled in the art will have access to that kind of power—that layer of Merina wrapping.

Speeches made by Merina elders fall into clearly decided patterns, following a fixed order and style of delivery including apologies, thanks, and blessings, as well as an abundance of proverbs and poems. The 'crucial proposal', which Bloch sees as the crux of the political matter, may emerge at the only point in the middle of all this where 'a certain freedom of expression' is allowed. This point was at first the entire focus of Bloch's attention, as a political anthropologist, but his paper, and indeed the whole book which it introduces, urge greater attention to the wrapping in which it is surrounded. The other papers describe a variety of situations in which the immediately evident power of influence and manipulation is contrasted with the less manifest power of formality and ritual skill.

This second type of power is not only less obvious, it is also often more difficult and time-consuming to acquire. In Japan, too, mothers are concerned with the manner in which their children say things because they want them to grow up to be polite, and they feel that this skill is most effectively passed on at an early age. At one level this operates to control the children. It also equips them with skills which they can eventually use to control others. Those who are able to be polite are also able to operate in a formal context, to deal with wrapping in its various forms.

This is an ability which confers considerable power on an individual, a power which may indeed be hidden from, or totally beyond the grasp of, other members of the society concerned. The best example I have seen in Japan—indeed, probably the single most influential factor in inspiring me to take up this project—was the headmistress of a kindergarten. She ran her establishment, including the teachers, parents, secretary, gardeners-cum-busdrivers, and even the visiting anthropologist, to say nothing of the children, with the most impeccable polite language. It was impeccably polite, and it kept everyone not only in their place, but constantly on their toes. As mentioned earlier, she was also concerned that everyone else should use polite language well, but she certainly set and maintained the standards.

In British society, too, the 'smooth-talk' a public-school education inculcates in a man is said to give him a presence which simply opens doors others are excluded by. There may be no very conscious realization that this is happening, but an encounter between people who recognize one another's evident *savoir-faire* in this respect, helps each to feel confident of the other's behaviour. It immediately inspires a trust, rightly or wrongly. This impression of respectability permeates down, as a neighbour of mine illustrated on several occasions when he 'sweet-talked' the helpers—police, psychiatrists, social workers—his distraught wife had called after noisy and violent drinking bouts he was prone to have. Unfortunately, the helpers would never appear until the next day, by which time the man had regained his immaculate Etonian wrapping.

As far as Japan is concerned, despite a commonly cited myth that 90 per cent of Japanese people consider themselves middle class, there are many levels of ability in the use of polite forms and other elements of *keigo* which also serve to identify subtle differences very similar to class distinctions in the British case. The distinct abilities are shared only by those who are conversant with specific speech forms, who are able to communicate in the 'restricted code' (Bernstein, 1974) they embody, and outsiders are to some extent excluded, whether they are aware of it or not. It is here that we enter the realm of the creation and maintenance of élites in society, and it was probably not unconnected to her concern with politeness that the headmistress cited above was seen by citizens unconnected with her kindergarten as engaged in the business of creating such an élite subculture in the town.

In Chapter 3 it was noted that those who are able to move between different levels of formality are in a particularly good position to manipulate situations, and thereby the people involved in them. Flexibility is an important skill to develop, and the practice of members of élite classes in Japan to drop the most elaborate speech forms in the company of those who cannot use them serves to wrap and protect the circles to which they belong, allowing these groups a form of relatively hidden power to communicate beyond the competence of outsiders. It also encourages a form of dissimulation, perhaps more appropriately described as the use of indirection.

Indirection and Dissimulation

This flexibility is to some extent developed by most Japanese speakers who learn from an early age to use different language in different situations. Users of local dialect in the home are encouraged to use standard Japanese at school, and they also learn polite forms of Japanese, possibly in a dialect variety, for ceremonial occasions within their own communities. Children brought up in homes which emphasize a high level of polite language equally learn to drop much of it amongst their schoolfellows. This practice of switching offers them some protection and a degree of power within their own local groups *vis-à-vis* members of the outside world. In some parts of Japan the local dialects were purposely developed to be incomprehensible to outsiders for protection from espionage during warring periods.

The ability to adjust runs alongside the necessity always to make distinctions between the front (*omote* or *tatemae*) appropriate for a particular situation and the real opinions (*ura/honne*) which lie behind it. The degree to which the *omote* deceives anyone depends on the skill of the speaker for, according to the psychiatrist and social commentator Takeo Doi, Japanese people very often know precisely what the *honne* is in any situation, despite the formal expressions of *tatemae*. He suggests that *omote* is precisely what expresses the *ura* that lies behind it so that when people look at *omote* they see also the *ura* through it. Indeed, they may be 'looking at *omote* solely in order to see *ura*' (1986: 25–6).

A clear confirmation of this idea is to be found in the examples in Chapter 3 of obviously untrue phrases, such as those proclaiming the worthlessness of an expensive gift, or that nothing is available to guests who have been invited to a sumptuous meal. On these occasions there is little misunderstanding. The phrases are part of the general expectations of dissimulation as part of politeness. Indeed, if the gift really were worth little, or the meal meagre, there would probably be much more indirect communication involved. This is not unknown, and the message would very likely be related to previous exchanges of a kind in some way unsatisfactory to the donor, unless perhaps an understanding of frugality existed between the parties concerned.

Sophisticated Japanese speakers are able to use several layers of *omote*, and though they may deny they have any intention to deceive, they are equipped to exercise the same kind of hidden power we discussed above. In the previous chapter it was pointed out that those most conversant with formal behaviour had most influence at meetings, and Moeran noted that an ability to hold one's drink and remember conversations held during the less formal parts of meetings, could be used for political ends (1984: 227). The more skilled communicators in this respect are undoubtedly able to convey information to a limited number of people, allowing others to understand only a part of their intentions, a complete version of which may be known only to themselves.[2]

This kind of indirect communication is by no means exclusive to Japan. In other areas, it may be more directly described as lying. In Greece, for example, where it is commonly said that 'knowledge is power' there is always a conscious effort to avoid revealing more than the minimum about a situation which could subsequently be used to one's own disadvantage, even if this involves telling direct lies. This practice has been described in some detail by du Boulay (1976), who relates it to the creation and maintenance of status and self-esteem in a situation where others constantly seek opportunities to mock and deflate these same efforts. A comparable situation, relating lying to a concern with the building up of an impressive self-image, has been described by Gilsenan (1976) for a community in the Lebanon.

Basso has written at length and with some eloquence about the practice of lying among the Amazonian Kalapalo people, whose deceit she describes as less to do with truth and falsehood than with the 'enactment of an illusionary relationship' (1987: 3). She argues that we must recognize, as the Kalapalo do, that our ability to lie creates opportunities and potencies as well as deficits and weaknesses (ibid. 242). Deception is probably the greatest source of humour here, and Basso argues that it allows resistance to coercion and the *status quo* (ibid. 356). An article by Howe and Sherzer (1986) about the people of San Blas Kuna makes a similar point about the way deception, used as a form of humour, allows the enforcement of egalitarian relations.

A monograph about the Awlad Ali Bedouin people, entitled *Veiled Sentiments*, is devoted to describing a rather different type of indirection, concerned with individual sentiments and the need to mask or 'veil' these in everyday life. Instead, they are expressed in poetry, and the author Abu-Lughod seeks to understand how 'individuals express such utterly different sentiments in poetic and nonpoetic discourse' (1988: 32). She relates the phenomenon to the values of the society, concluding that the poetry can 'be viewed as their corrective to an obsession with morality and an overzealous adherence to the ideology of honor' (ibid. 259).

In a volume of essays about Pacific communities, Brenneis and Myers (1984) have gathered together a number of papers which relate indirection of one sort or another to the political situation of the people concerned. Comparable with Howe and Sherzer's material above is the Fiji case, discussed by Brenneis (1984), who has elsewhere made a direct comparison with the Caribbean (1987), where he argues that, in both cases, indirection is used as a way to deal with problems of local egalitarianism in a wider system of hierarchy. In Fiji, he points out that it resolves the apparently contradictory requirements that one must 'both act politically and avoid the appearance of such action', noting that 'the perils of direct leadership and confrontation in such societies often foster indirect, metaphoric and highly illusive speech' (1984: 70).

In another paper in Brenneis and Myers's volume, Atkinson uses the wrapping metaphor in a discussion of the importance of 'wrapped words', among the Wana people of Sulawesi, to 'hint at a meaning without confronting others directly' (1984: 43). Her paper describes a highly formalized type of speech, or poetry, which is used to express elegant, but nevertheless very often political, statements in an indirect way, disguised from all but those who share knowledge of the appropriate context. She expresses the sense that this form of language is 'in keeping with a cultural avoidance of confrontation and open conflict' (ibid. 42), and explains that politics among the Wana is concerned with attracting followers rather than coercing them, an idiom more appropriately courtship than *realpolitik* (ibid. 60).

In both these cases we are concerned with political activity which is not immediately obvious, sometimes even to those involved. We are also concerned with a cultural preference for the avoidance of conflict.[3] These are excellent examples, with parallels in Japan, of the argument of Brenneis and Myers, reiterating the point made above, that we sometimes need to look further for the locus of power than the politics of our own societies would suggest. We need to see, for example, that 'politics is concerned not only with exercising power but also with reproducing the mechanisms which make power possible' (1984: 4).

These authors praise Bloch's (1975) work, discussed above, for drawing attention to the importance of language and the 'hidden' controls it may embody—for the way, like Bernstein (1974), he identifies power and control 'that is permeated through social intercourse in a totally unconscious and completely accepted way' (Brenneis and Myers, 1984: 9). However, they also notice a confusion in Bloch's argument and criticize his apparent unwillingness, an unwillingness characteristic of a transactionalist approach, to move beyond the level of action and individual strategy to discuss the importance of oratory for the reproduction of structures of social relations. In fact this role of particularly the more formal types of oratory clearly emerges in several of the papers in both books.

Back to Material Objects

In Japan, a nation well known for the value it attributes to harmony and the avoidance of open conflict, emphasis is placed not only on the power forms of speech may convey, but also on the importance of non-verbal communication. Just as space is important in Japanese paintings, and other forms of art, the spaces in conversation are also said to be vital to obtain a deep understanding of what is being communicated. Often what is not said is just as important as what is said. This point was made forcibly at the end of Chapter 7 in the discussion of the ritual of the tea ceremony. The actual speech used is very sparse, extremely restricted, and we would learn very little if we looked only at the words.

In general, in the Western industrialized world, we place a great deal of emphasis on verbal means of communication and perhaps even more on its printed version in the form of books, newspapers, learned papers, and so on. The ability to read and understand difficult texts, and to write skilfully, confers certain forms of power, partly again because of the time, effort, and to some extent money, required to acquire those skills. Similar points were made in an interesting article about the case of the power of Arabic literacy, and its possibilities for 'obfuscation, concealment of meaning and lying', among the Mende of Sierra Leone. The authors partly explained the way Arabic writing was thought so powerful because it is used primarily for ritual purposes (Bledsoe and Robey, 1986: 222).

In Japan, too, literacy is valued, indeed it is an almost universal skill, but to understand the proverbial space between the lines we might more profitably look to some of the other societies with which Japan has been shown to have similarities. Our emphasis on articulating relations of power may make it harder for us to recognize other aspects of non-verbal communication, ritual and otherwise. So far in this chapter we have been concerned mostly with language, at the level of oratory and speech-making. Bloch noted the emphasis during socialization on the manner in which things were said, on gesture and body position, but he may not have gone far enough. In an interesting article about cloth and the wrapping of ancestors to tap their power in another part of Madagascar, Feeley-Harnik also suggests that Bloch may have overlooked a layer of (this time literal) wrapping (1989: 102).

In Chapter 2 we mentioned Weiner's paper about the use of yams in the Trobriand Islands as an indirect form of communication, and this offers an alternative vehicle for the transmission of ideas, as well as the exercise of power. Amongst a people with whom it is taboo to mention directly negative feelings one may harbour for them, the yam harvest is a time for the non-verbal expression of such opinions. In other words, on this again somewhat ritualized occasion, the quantity of yams picked, when people help one another with the task, may serve as an indication of underlying content or discontent (Weiner, 1983).

Another example mentioned in Chapter 2 was Guss's (1989) study of the Yekuana people of Venezuela where he discovered how much more knowledge had become available to him when he began to engage in the making of baskets. Here, too, the activities associated with collecting and preparing the materials, rituals made necessary by the removal of these items from the environment, discussions about the type and value of different designs, and the weaving process itself, all helped to make Guss aware of means of communication beyond the narrative he had set out in search of. Guss found out about the creation myth he was after, but he learned a lot more about the transmission of knowledge in the process.

Michael O'Hanlon's (1989) study of the Wahgi people—*Reading the Skin*—is a detailed exposition of the way economic and political matters can be properly understood only by incorporating indigenous notions of communication through material objects such as the elaborate wigs worn, again on ritual occasions. However, in a more recent paper (1992) about a particular kind of wig, he suggests that expectations about kinds of 'talk' may have blinded observers even to aspects of verbal communication. In an area of New Guinea well known for its lack of verbal exegesis, O'Hanlon argues that people do talk about the wigs, but they do it by making assessments, assessments which 'index an indigenous theory of significance, in which outer forms are felt to monitor and authenticate otherwise concealed states and processes' (O'Hanlon, 1992).

Particularly relevant to this chapter is Howard Morphy's interpretation of the art of the Yolngu aboriginal people of northern Australia. Yolngu art, he explains, is not only a system of encoding meaning about their ancestral past, and therefore part of the general system of religious knowledge, it is also part of a system of restricted knowledge, access to which confers power to various categories of people. Sometimes these people gain such access according to their gender, and/or at various stages of the life cycle, but an overall understanding of the system, which he describes as an 'apparent layering of knowledge', depends on an understanding of the indigenous concepts of inside and outside. This distinction is described as an 'all-pervasive' aspect of Yolngu culture (1992: 75–8).

It is not possible to go into much detail here about the Yolngu system of knowledge because the point is precisely that a full understanding of it is embedded in other aspects of the culture. However, there are some interesting parallels with the Japanese case. The distinction between inside and outside is also crucial in understanding Japanese society (Hendry, 1987), and there are some similarities in the way these concepts are perceived. The notions of *omote* and *ura* (translated as 'front' and 'rear') would apply well to an example, given by Morphy, where the aboriginal people objected to the construction of a road, proposed by the Australian government, because the government representatives only gave them 'outside' reasons of why they thought the road would be good for the Yolngu people, but had neglected to explain the 'inside story' about why they wanted it for themselves (1992: 81). An understanding of this objection is surely not limited to Japan and Yolngu!

The notion of layering in the Yolngu case is also reminiscent of the layers of Japanese wrapping we have described, and the way sophisticated Japanese speakers are able to use several layers of *omote*, as discussed above. Morphy also refers to Frederick Barth's depiction of ritual knowledge among the Baktaman people of New Guinea as 'parceled out into a series of packages' (ibid. 92), and Condominas has described Thai political systems as a *système à emboîtement* (Condominas, 1978: 106). The important point is that knowledge in any society, and the power with which it is associated, is organized in a way which may be peculiar to that society, with a specific rationale which may be almost inextricably involved with the material culture of that society, but there are also points of contact. With an open mind, and a degree of sympathetic imagination, communication is certainly not impossible.

We are beginning to come full circle now, back to the ideas mooted in the

Introduction to the book, ideas that I have tried to elaborate in each of the chapters in turn. In Chapters 1 and 2, for example, we considered in some detail the communication possible through various ways of presenting gifts; in Chapter 4 we saw the hidden power carried by disarmingly gauche, but well-dressed, Japanese businessmen. A more sinister power, hidden one layer underneath such suits, in the tattoos of Japanese gangsters, is said to inspire considerable fear and respect, especially within a prison community.

The fact that the tattoos are usually hidden offers an interesting material expression of the way organized crime operates in Japan. It is hidden under front occupations, such as newspaper publishing, but just as everyone is aware that the *omote* in language is only a front for the *ura* which lies behind, most people, including the police, are aware of the underworld as well. The low crime statistics published by Japan are another *omote*, and Japanese people are well aware of this, but they make so little of the organized activities which help to maintain it that an academic criminologist spent several months there without becoming aware of its existence (Fenwick, 1989, and personal communication). In practice, as is occasionally witnessed when scandals erupt the Japanese *yakuza* underworld is extremely powerful, both in Japan and in several other countries of the world (see e.g. Ames, 1981; Kaplan and Dubro, 1986).

This is a striking example of the kind of hidden power we have been gradually identifying, emphasizing the importance of looking beyond the most immediately evident aspects of an apparently political situation. In Chapter 5 we also talked about buildings which have deceptive exteriors, or exteriors which can only be accurately assessed by those in possession of considerable inside knowledge of the particular cultural forms. We also talked about the need to understand culturally variable ways of behaving within certain types of space. Those who are not only aware of, but also at ease in, specific cultural situations, are evidently able to exercise more control and power than those who are struggling to observe the rules.

In the Japanese case I have tried to demonstrate parallels in the various different arenas discussed. I have illustrated the propensity to take trouble with 'wrapping' from the prototype gift to language, garments, architecture, time—even the way humans arrange themselves. The value-laden ideas of refinement and civilization associated with the type and layering involved would appear to run through these different expressions of wrapping forms, culminating in perhaps the ultimate expression of Japanese art, which brings so many forms together, namely the tea ceremony.

The ability to create and adjust such wrapping must, I suggest, operate a crucial influence in the exercise of power at any of these levels, and, in practice, several levels will operate together. The language of people dressed in formal kimono is noticeably more polite than that of the same people dressed in sports gear, and a beautifully wrapped gift cannot just be thrown down on the doorstep. At all levels, and between levels, an awareness of and ease with appropriate behaviour leaves options open to use and manipulate it in ways which may barely be recognized by those with less facility.

This differential access to power, related in particular to knowledge about material objects, is similar to that described by Douglas and Isherwood in their attempt to apply an anthropological approach to the study of consumption behaviour. They suggest, from the ethnography they gather, that, 'regardless of how evenly access to the physical means of production may be distributed, and regardless of free educational opportunities, consumers will tend to create exclusive inner circles' (1979: 180–1). They argue that people will try to gain access to the top consumption class largely because of the higher earnings this will eventually make available to them (ibid. 181–3).

In fact, when we become this materialistic, I think we are beginning to be more concerned with status than with power, a point demonstrated by taking a last comparative look at the very top of these 'top classes', in some of the Western countries considered by Douglas and Isherwood, and in Japan. If we limit ourselves to European countries with monarchs, we have a reasonable comparative base for Japan, for the people who stand at the pinnacle of these societies have access to a tremendous range of goods and services, usually unthinkable for most of their subjects.

If their lives are measured against the wants and needs of those ordinary people, they are seen to be in possession of enormous amounts of wealth and, within certain limits, they are able to spend a good proportion of their lives engaged in pursuits which others can only admire from a distance. When they visit one another their facility with the formal exigencies of the wrapping within which they conduct their lives undoubtedly gives them a means of communication beyond that of their mutually incompatible languages, although these people are of course subject to misunderstanding of the familiar as is every other member of their respective societies. Nevertheless, these are people who unquestionably have access to a great deal of status and material power.

Of course, power and its strength depend in the end on what people want. Status is for most a very important element, and its acquisition alters the nature of every new relationship as it is subsequently redefined. If it is accompanied by the ability to operate appropriately in a formal situation, it may be particularly potent. This may be described as a kind of ritual power, which should certainly not be dismissed as therefore empty. It may automatically bring others to work on behalf of a cause one supports, a benefit, even if it only allows one to allocate time to something else one wants to do. If this can be accomplished in such a way that those others feel they are themselves in control of their own lives, perhaps even those of others, so much the better.

Some of the people who have most status in society are also said to have very little power, however. Bloch uses a British metaphor when he describes 'the carriers of the formal code' as having been 'kicked upstairs' because of the gradual impotence which it causes (1975: 28). Members of the British royal family, like members of the Japanese imperial family, are in some ways virtual prisoners within the wrapped spaces in which they reside. They are unable to go out into the world of ordinary people without the protection of a fair measure of wrapping of one sort or another, social and/or physical. In other words, the price they pay for their lifestyle is the freedom to operate as others do.

The real power of these royals, then, is, ironically, hardly available to them. Indeed, their wrapping serves to protect them in person from the power of their positions, which exists largely to be harnessed by others. In the case of Japan, the emperor stands for the people, and his presence is appealed to as a symbol of the Japanese nation. He may not have much power to influence people but his very existence gives power to others. Before and during the Second World War the Japanese people were assured of his divinity, and understood that they were all also descended, at whatever remove, from the same ultimate ancestor. In his name they undertook to fight for the nation, and in the most devastating of circumstances they pressed on with their mission, even if it meant dying for this man they would never see.

He has formally renounced his divinity, but some of that wrapped power is still available. The extraordinary concern with the death of the Showa Emperor, and the international presence at his funeral and the enthronement of his successor, express the way he stands for a now very powerful people. It is because of this continuing symbolic power that people in Japan and the United Kingdom demonstrate against him from time to time, though for different reasons. In the first case it is usually because he represents a system they want to do away with, he stands for all that they see as feudal and outmoded. In the second it is because he symbolized the atrocities they saw as committed in his name.

The extent to which the person of the Emperor Hirohito was involved in decision-making about events before and during the Second World War, has consumed some journalists and historians recently, but I feel their interest in this subject is a somewhat misplaced result of their own biased understanding of politics. In the so-called democratic world in which we live, until recently largely dominated by the United States, real and symbolic power are seen by many as almost synonymous. We should be careful not to overlook the layer of wrapping which in Japan, and some countries of Europe, is still sometimes quite powerful.

Wrapping it Up
The Symbolic Power of Wrapping

The kind of power we have been discussing here is largely what may be called symbolic, or ritual power. In view of the various forms of wrapping we have discussed, this power is evidently diffused very effectively throughout social life, so that it is difficult at a particular time to identify precisely where it lies. Moreover, it may lie in several places, each reinforcing the others. It is thus perhaps the kind of power described by Bourdieu as 'that invisible power which can be exercised only with the complicity of those who do not want to know that they are subject to it or even that they themselves exercise it' (1991: 164).

The importance of the layers of refinement which wrapping represents in Japanese society is evident in the way they somehow persist through huge influxes of outside influence. There may be modifications, even large-scale transformations, but the basic underlying principles seem to remain. There may have been much talk of conservation and the need to cut down on the apparently extravagant use of layer upon layer of paper and other forms of wrapping, but as yet the only evidence that this is happening in Japan would seem to be found in recycling efforts. A shop has even opened in London, designed especially for Japanese tourists, where British goods are sold in the style the Japanese customer is used to—well wrapped, and handed over politely.

At the start of this book, I suggested that the range of uses of the word 'wrapping' may be regarded just as metaphorical expressions, as analogues of the paradigm posed in the use of objects, notably gifts. This is a minimalist position, however, and I hope I have shown that the notion of wrapping is more deep seated than this. The 'wrapping principle', as I used to describe it in the early days of this investigation (Hendry, 1990*a* and *b*), has proved to be a most pervasive part of Japanese life. Indeed, even as I write the closing pages of this book, my colleagues are still coming up with new examples.

In some ways it is a 'logical schema', like the concepts of inside and outside which 'provide a rationale for the existence of levels of knowledge' among the Yolngu (Morphy, 1992: 78). In this sense, as Morphy notes, it could perhaps be described as a 'diagram' in the way that Parmentier used the term to analyse the myth and history of Belau. Parmentier makes the apt point that, from the Belauan point of view, such 'prototypes are not merely convenient objects for metaphorical or analogical reasoning, but are in fact presupposed causal models which generate patterned social reality' (1987: 113). The wrapping model could certainly be described as part of a 'structuring structure', as Bourdieu describes 'instruments for knowing and constructing the objective world' (1991: 165).

In Geertzian terms, the wrapping model is simultaneously a model of, and a model for social reality (1975: 93). James Valentine has already cogently applied this idea to a type of Japanese dance known as *nihon buyō*. He sees the delimited physical space drawn by women's movements in the dance as symbolic of their restricted cultural space, a model of women's social role, at the same time acting as a model for women, 'reinforcing restraint and self-control'. Indeed, several of his informants had taken up the practice precisely to engender these qualities, 'control of both body and emotions for the sake of an elegant humility' (1986: 124). Here is yet another example of Japanese wrapping to add to the list, an example which nicely breaks down the material/non-material divide we are prone to make.

Wrapping in Japan is a veritable 'cultural template', or perhaps we could add another metaphor and call it a 'cultural design'. It makes possible the marking of the whole range of life-stages and statuses, thus representing, and recreating, the hierarchical order which, in turn, gives rise to the locus of power relationships. Different manifestations of this organizing principle reflect and reinforce one another, as has been shown at various points in the book, and they thus also offer almost unlimited possibilities for communication, verbal and non-verbal, and for the exercise of power.

The value of the metaphorical uses of the idea of wrapping and packaging is not limited to Japan,[1] however, as we have seen in various chapters, and as we are reminded in everyday conversations, at least in the English language. In all societies, we use various forms of wrapping to express our opinions, and to exercise our powers of persuasion, whether we use that metaphor to describe them or not. In the last chapter we examined a wide range of what could perhaps be called 'political packaging types', exercising different forms of wrapping.

The relationship between different manifestations of wrapping will be different in different societies, however, and we have here only been able to examine in detail the albeit enormous potential of this one structuring form in the case of Japan. Nevertheless, each society will have its own diagrams, logical schema, or cultural templates, and it is of course up to each ethnographer to identify those appropriate to their own peoples. One anthropologist who heard a paper I gave on the subject of wrapping began to muse about the notion of winding being useful in thinking about India—the turban, the sari, folded books, and the absence of a notion of inside and outside in several contexts.

This is not the place to start hypothesizing about other cultural designs, however. It may be that these are manifest in quite different ways anyway. We have learned a lot here by comparing Japan with a wide range of other societies at various different stages of technological achievement, and in view of the clear cross-cultural value of the notion, there is no reason why it should not be possible for people other than anthropologists to seek a greater understanding of the wrappings of still more people, just as the Japanese have travelled around the world examining our wrappings and those of others like us.

It is evidently important not to try to take off the layers of wrapping we find elsewhere, always to be seeking essences, because in this way we may be throwing out some of the important cultural information we need, perhaps only to find nothing at all—or a strange, significant emptiness—inside. We must try instead to examine our own wrappings, so that we can identify our own prejudices, and not allow these to blind us to an understanding of those of others. It may well be harder when these are similar, as in the case of gifts, which we also like wrapped, than for a people who are suspicious of things that are too elegantly enveloped. However, a look at this one cultural design may now help us to know where to look for the locus of power in situations we might not otherwise have identified.

At its most complex, this is of course the whole work of anthropologists, and we have identified nothing more than the subject-matter of our field. On the other hand, I would like to think I have also opened up some new vistas for the traveller and the international business executive of the jet-setting generation. Fortunately, for the most part, some of these new travellers do seem to be more sympathetic than their great-grandparents were to different ways of thinking. Let us hope they will help us to avoid being 'Japanned' in this area as well as in an economic one!

NOTES

Introduction

1. My use of the word 'arena' is not as precise as that of Turner (1974), or his predecessors, although it is sometimes describing the location of antagonistic interaction. The subjects of different chapters of the book give a good indication of the range of meaning implied.
2. I am grateful to Michael O'Hanlon for reminding me of this aspect of Mauss's argument. The point I am making here is also not unlike that made by Daryl Feil in his insistence on a consideration of the material side of exchange transactions in his paper entitled 'From Pigs to Pearlshells' (1982: 291), although the subsequent analysis has a different focus.
3. The quotation, from Nohara Komakichi, *The True Face of Japan*, was cited by Benedict (1954: 290). I am grateful to Robert Smith for pointing the parallel out to me.
4. An apt illustration of this value system is that whereas academic publishers will pay for the words of anthropological books, they very often insist on outside support to help cover the cost of photographs, if they use them at all. Oxford University Press must be thanked for agreeing to carry the photographs in this book, although, as acknowledged earlier, the Daiwa Foundation covered the bulk of the expense.
5. Geertz points to the criticisms of the genre made by people who he imagines saying 'What we want to know about is the Tikopians and the Tallensi, not the narrative strategies of Raymond Firth or the rhetorical machinery of Meyer Fortes' (1988: 1–2), but he himself argues that the ability of anthropologists to get others to take them seriously is most of all concerned with 'their capacity to convince us that what they say is a result of their having actually penetrated (or, if you prefer, been penetrated by) another form of life' (ibid. 4).

Chapter 1

1. The way these cosy-sounding names undoubtedly help the acceptance of prepared foods, can plausibly be compared to the way personalization helps sell commodities at least partly through the attachment to them of the symbolism of 'possession', an idea discussed recently by Carrier, who incidentally uses the interesting notion that the objects are 'wrapped in this symbolism' of possession (1990: 702).
2. It is not unheard of for people to save wrapping paper, in fact I do so myself, although I could be said to have a professional interest in the stuff. I do also know someone who irons it ready to use again!
3. Cobbi points out in an article on the obligation of the gift in Japan that although the word *cadeau* in French implies a notion of pleasure, this is not generally the case in Japanese (1988: 113).
4. A 'graph' of such a breakdown, with illustrations of the various, increasingly expensive gifts exchanged, is presented in Morsbach (1977: 102).

5. According to one Japanese source, the name *noshi* comes from the verb *nosu* (to stretch) because the abalone is stretched through a process of steaming and beating. The samurai apparently chose this as a symbol of celebration because the word for 'beat' (*utsu*) is homophonous with the word for 'defeat' (Ema, 1971: 111, 224).
6. According to one book of etiquette, *mizuhiki* was originally imported to Japan from Tang China, and it carries a meaning of expressing respect for the recipient (Ogasawara, 1985: 7).
7. This phrase sounds as though it could warrant a consideration of the possible spiritual content of a gift, as discussed by Mauss (1954) in reference to Maori and Samoan gifts, and later taken up by other commentators (e.g. Parry, 1986), but there would not seem to be much relevance to the argument at this point.
8. In view of the Hindu notions about the possibilities of discarding inauspiciousness through the giving of gifts (Parry, 1986; Raheja, 1988), this custom could represent a Shinto purification of an imported Buddhist custom, but I have as yet no evidence for this idea.

Chapter 2

1. I am grateful to Mike O'Hanlon for pointing out a parallel here with the dual role of money in New Guinea societies, sometimes used as a valuable to be presented to people, but also used as currency (Healey, 1990: 200–2).
2. These groups are discussed in more detail in English in Hendry (1981*b*), and in much greater detail in Japanese in Sakurai (1962).
3. In fact standard souvenirs of particular parts of Japan can be purchased at major railway stations and airports, so that an omission may in practice be rectified at the last minute, but many people will take trouble to seek local products which are known only to be available in their place of manufacture. This kind of knowledge is a rather typical part of the way people are prone to assess gifts they are given, not always charitably.
4. Kyburz (1988: 26) actually refers to 'the cashmere scarf of Burberry's, easily identified as an English product by its tartan pattern', a strange sentence which perhaps represents accurately a Japanese view of the message, but which neglects cultural relationships which have gone into the creation of this 'English' symbol. These are, of course, the fact that tartan is more often associated with Scotland than with England, although Burberry has a very distinctive pattern of its own, and that the very name 'cashmere' betrays a former British link with India.

Chapter 3

1. The translation of the original Japanese word, *sui*, as 'essence' here actually does not imply any of the meaning of 'essence' as discussed in other parts of this book as something which Westerners are concerned to reveal. Other possible translations of the word, apart from 'cream' are 'purity, pith, pick, élite, choice, elegance, gracefulness' (Nelson, 1974), 'quintessence and genius' (Kenkyūsha, 1954).
2. Lola Martinez reports (personal communication) that in Kuzaki, in Mie-ken, where she carried out fieldwork, the villagers assumed from watching dubbed

television films that the English translation of this word was nothing more than 'darling'.

3. There is a recognized type of rudeness which involves overpoliteness in Japanese. It is known as *ingin burei*, literally 'polite impoliteness', or 'polite rudeness'.
4. Further details of all these rites can be found in Hendry (1981*a*: ch. 5).
5. The word used to describe tea emerging from a pot—*deru*—is the same word which is used to describe a bride leaving home, and since poor tea can only be served once, as opposed to good tea which can bear several additions of hot water, this symbolizes that the bride should only leave home once.
6. Wrapped presents may also be placed in the Buddhist altar for a while before they are opened, or when they have been opened only slightly, but I think this is a sharing of gifts with the ancestors, rather than an offering *per se*. Since my research would suggest that, in the case of gifts, the wrapping is at least as important as the present inside, it would seem appropriate to place the gift in the altar in its wrapped state.
7. Uno quotes some letters to newspapers on the subject (1985: 17–21).
8. A passage in *The Buddha Tree* (Niwa, 1966), emphasizes the value accorded to a priest because of his chanting voice, which is argued to justify his continuing practice in a community despite a scandalous affair with one of his parishioners.
9. A similar contrast between form and content is made by Bachnik (1986: 65–6) in comparing the actions of the same participants in a formal ritual scene with those of an informal kitchen scene. She also notes the importance of the aesthetics of the ritual scene.
10. The overall thrust of Bloch's argument is to do with the implicit power associated with ritual (see Ch. 8 for a development of these ideas). The recent boom of *karaoke* machines, and their associated ritualized singing, could perhaps be cited as an interesting implicit expression of Japanese economic power.
11. The extent to which the power of words is related to their incomprehensibility, and therefore perhaps to their secrecy, was discussed some time ago by Tambiah (1968). His argument illustrated the advantages of a textual analysis in the case of the Trobriand Islanders, but the pronunciation of the original Sanskrit of these Buddhist texts has now been so distorted that only a scholar of the language shifts involved would be likely to be able to extract much meaning.

Chapter 4

1. Gillian Feeley-Harnik (1989) uses this analogy to good effect for the clothing of the Malagasy people of Madagascar, however, particularly in reference to the wrapping of the ancestors in human form.
2. In view of this argument, I am slightly perplexed about how to interpret some of the extraordinary slogans which appear on modern Japanese T-shirts. These are often in English, a language which most of their wearers will have spent several years studying, yet they also often display awful grammatical errors as well as quite crude direct meanings (see e.g. Moeran, 1989: 1). Comments on these slogans generally reveal a blissful unawareness of their literal meaning, and I can only assume that there is a considerable lack of subtlety here, so that the garments merely look

(attractively) Western. Sherry and Camargo have offered an analysis of the English-language labelling on Japanese beverage cans, and there may be something in their point that 'Focusing on what a word actually means seems more a Western than a Japanese disposition' (1987: 179).

3. A measure of the seriousness taken in adopting Western dress is indicated in books of etiquette which may even advise on how to walk in high- and low-heel shoes (see reproduced page in Rudofsky, 1972: 195).
4. As long ago as 1920 a Turkish writer pondered the way the Japanese manage to retain their religion and culture while entering successfully into 'Western civilization', exemplified at the time by their acceptance (and not Turkey's) into the League of Nations (Gökalp, 1963: 58).
5. Rudofsky has reported a nice reversal of 'our concepts of modesty' when he notes that 'in some parts of the world only harlots wear clothes' (1972: 26).

Chapter 5

1. According to the Japanese historian of architecture Terunobu Fujimori, the idea of architecture as 'the foremost monument of history and civilization' is an entirely Western one. In contrast, he claims, 'Japanese have never thought of architecture as anything but temporary and transient' (1990: 15). However, since in the same article he also claims that some Japanese temples are 'nearly two thousand years old' (ibid. 11), I find his argument a little unconvincing. The way the structure of famous Japanese shrines and temples is regularly renewed, even employing hereditary carpenters in the vicinity for the purpose (Coaldrake, 1990: 16), would also seem to contradict that view.
2. This clear division in design dates back at least to the samurai house of the Edo period where the reception part and the family part of houses were 'strictly separated from each other' (Yoshida, 1955: 29; cf. M. Inoue, 1985: 108). Pezeu-Masabuau (1981: 53) also notes that the *zashiki* was sometimes referred to as *dei* or *omote* (*devant*), with the direct meaning of being the front or outside of the house.
3. Pezeu-Masabuau commented on the different behaviour required as one moved 'up' and 'down' between different levels of a Japanese house, noting in particular the change of floor surface, from *doma*, or baked earth, which was usual in the kitchen and most *oku* regions, through *itanoma*, or wooden boarding found in intermediate areas (and, more recently, in kitchens), to the *tatami* surface of rooms such as the *zashiki*. He suggests a connection here with strong notions of social constraint and hierarchy with the use of the house space (1981: 444–5).
4. This is a subject which has been developed extensively by W. H. Coaldrake (forthcoming).
5. This is apparently even more true than Barthes may have realized, for an acquaintance of mine who was invited to stay in the palace by a member of the imperial family to whom she had become close at Cambridge, found it almost impossible to persuade taxi-drivers to take her further than the gate, a substantial walk from the place where she was lodged!
6. Some detail about the different types of wood used is to be found in Coaldrake (1990: 20–3).

7. It is true that a strange man did once make his way into the palace and surprise the Queen in her bedroom, but the incident caused much amazement in the world at large. The important thing is that although the palace is visible, it also appears to be impenetrable.
8. For further discussion about the importance of lying in Middle Eastern communities see e.g. Gilsenan (1976) or Asad (1970).

Chapter 6

1. More detail about how this dropping of the layers takes place will emerge in Ch. 7.
2. The almost morbid interest in the failing bodily functions of Emperor Hirohito may seem to prove the opposite of this statement, but it doesn't counteract the fact that this is the way people conceive of the Emperor. The way in which people grew tired of the length of the uncertain situation when the Emperor was about to die would also suggest that the concern was protocol rather than genuine sorrow for the man himself.
3. This idea was apparently also found in prisons, at least in the period preceding Westernization, where *tatami* would be stacked up to provide a place of honour for the bosses (Fujimori, 1990: 13).
4. See Gow (1988: 26) for some examples of titles, and some of the difficulties of translation.
5. Some further detail about the structure and workings of Japanese companies can be found in Kono (1984), and Abegglen and Stock (1985). More practical advice for those doing business in Japan is available in Fields (1983), Zimmermann (1985), and Morgan and Morgan (1991).

Chapter 7

1. Vogel (1975) is a good collection of essays on the subject of decision-making and other aspects of organization in Japan.
2. Another detailed study of the internal dynamics of a drinking occasion, this time held in an urban context, is in considerable agreement with Moeran's analysis (Ben-Ari, forthcoming).
3. Indeed, I have argued elsewhere (Hendry, 1991) that this degree of self-determination in the lives of Japanese schoolchildren helps to foster positive attitudes towards education.
4. An extreme version of this practice is the process of self-mummification whereby an ascetic would gradually discontinue various parts of his diet until his body literally dried up. His skin would then apparently resist the usual rotting process and his devotion would represent a great deal of spiritual power. Sometimes these practitioners would ask to be buried before they actually died, but it seems that more often they would die before reaching their goal (Blacker, 1975: 87–91).
5. Anderson (1987: 483) gives further detail about the factors which influence these choices, describing the decision-making process as 'a series of delicate discriminations ritually locating the gathering within geographic, religious and ethnic boundaries as well as establishing cyclical and historic time coordinates'.

Chapter 8

1. Irvine has discussed various ways in which the terms formal and informal have been rather indiscriminately used. I am aware of this discussion, but feel happy at present to use the words as 'cover terms' (1979: 785) which imply some aspects of each of her uses. I think her argument about the way 'formal' meetings may be distinguished from the decision-making process, which may or may not be part of these meetings, is particularly valuable.
2. This type of indirect communication was also discussed long ago by Evans-Pritchard in reference to the use of *sanza* among the Azande (1962: 228). He emphasized the problems for the ethnographer of trying to understand things which only a limited number of native informants could follow, and noted the importance of relating such practices to the context of the wider social system in which they are found.
3. Strathern's discussion of 'veiled speech' in Mount Hagen is another example of the use of indirection in the avoidance of open conflict, even, in this case, outright violence (1975: 199). In another study of the language of the New Guinea Highlands, Goldman considers the relationship between form and meaning by looking at aesthetic and rhetorical norms for speech used in disputes (1983: ch. 5).

Wrapping It Up

1. Just as this book is about to go to press Michael O'Hanlon has brought to my attention some of the content of another book in press in the same series, namely *Wrapping in Images*, by Alfred Gell. There would seem to be several similarities with Japan in the Marquesan Islands, which is the focus of his research, even beyond the full-body tattoos. For example, wrapping is there too concerned with keeping the sacred apart from the mundane and the notion can be applied to a chief's retinues of servants.

BIBLIOGRAPHY

ABEGGLEN, JAMES, and STOCK, GEORGE (1985). *Kaisha: The Japanese Corporation*. New York: Basic Books.

ABU-LUGHOD, LILA (1988). *Veiled Sentiments: Honor and Poetry in a Bedouin Society*. Berkeley: Univ. of California Press.

ALEX, WILLIAM (1963). *Japanese Architecture*. New York: George Braziller.

ALLYN, JOHN (1970). *The 47 Rōnin Story*. Rutland, Vt.: Tuttle.

AMES, WALTER L. (1981). *Police and Community in Japan*. Berkeley: Univ. of California Press.

ANDERSON, JENNIFER L. (1987). 'Japanese Tea Ritual: Religion in Practice', *Man*, n.s., 22: 475–98.

ARAKI, HIROYUKI (1983). *Keigo Nihonjinron*. Kyoto: PHP Kenkyūjo.

ARAKI, MAKIO (1978). *Tezukuri no Kurashi: Orikata-Tsutsumu Kokoro*. Tokyo: Bunka shuppan kyoku.

ASAD, TALAL (1970). *The Kababish Arabs: Power, Authority and Consensus in a Nomadic Tribe*. London: C. Hurst.

ATKINSON, J. M. (1984). 'Wrapped Words: Poetry and Politics among the Wana of Central Sulawesi, Indonesia', in Brenneis and Myers (eds.): 33–68.

BACHNIK, JANE M. (1986). 'Time, Space and Person in Japanese Relationships', in Hendry and Webber (eds.): 49–75.

BAIZERMAN, SUZANNE (1991). 'The Jewish *Kippa Sruga* and the Social Construction of Gender in Israel', in Barnes and Eicher (eds.): 92–105.

BARNES, R. H., DE COPPET, DANIEL and PARKIN, R. J. (1985). *Contexts and Levels*. Journal of the Anthropological Society of Oxford, Occasional Papers, No. 4. Oxford.

BARNES, RUTH, and EICHER, JOANNE B. (eds.) (1991). *Dress and Gender: Making and Meaning in Cultural Contexts*. New York: Berg.

BARTHES, ROLAND (1970). *L'Empire des signes*. Paris: Flammarion.

—— (1982). *The Empire of Signs*, trans. Richard Howard. London: Jonathan Cape.

BASSO, ELLEN B. (1987). *In Favour of Deceit: A Study of Tricksters in an Amazonian Society*. Tucson, Ariz.: Univ. of Arizona Press.

BEFU, HARUMI (1966). 'Gift-Giving and Social Reciprocity in Japan', *France-Asie*, 188: 161–77.

—— (1968). 'Gift-Giving in a Modernizing Japan', *Monumenta Nipponica*, 23: 445–56.

BEIDELMAN, T. O. (ed.) (1971). *The Translation of Culture*. London: Tavistock.

BEILLEVAIRE, PATRICK (1986). 'Spatial Characterization of Human Temporality in the Ryūkyūs', in Hendry and Webber (eds.): 76–87.

BEN-ARI, EYAL (1989). 'At the Interstices: Drinking, Management and Temporary Groups in a Local Japanese Organization', *Social Analysis*, 26: 46–65.

BENEDICT, RUTH (1954). *The Chrysanthemum and the Sword*. Tokyo: Tuttle.

BERNSTEIN, BASIL (1974). *Class Codes and Control*, i. London: Routledge & Kegan Paul.

BERQUE, AUGUSTIN (1990). 'The Rituals of Urbanity: Temporal Forms and Spatial Forms in Japanese and French Cities', paper presented at the Japan Anthropology Workshop, Leiden, 1990.
BESTOR, THEODORE C. (1989). *Neighbourhood Tokyo*. Stanford: Stanford Univ. Press.
BLACKER, CARMEN (1975). *The Catalpa Bow*. London: Allen & Unwin.
—— (1990). 'The Shinza or God-Seat in the Daijōsai—Throne, Bed, or Incubation Couch?', *Japanese Journal of Religious Studies*, 17/2–3: 179–97.
BLEDSOE, C. and ROBEY, K. M. (1986). 'Arabic Literacy and Secrecy among the Mende of Sierra Leone', *Man*, n.s., 21: 201–26.
BLOCH, MAURICE (1974). 'Symbols, Song, Dance and Features of Articulation: Is Religion an Extreme Form of Traditional Authority?', *Archives européenes de sociologie*, 15: 55–81.
—— (1992). *Prey into Hunter: The Politics of Religious Experience*. Cambridge: Cambridge Univ. Press.
—— (ed.) (1975). *Political Language and Oratory in Traditional Society*. London: Academic Press.
BOGNAR, BOTOND (1985). *Contemporary Japanese Architecture*. New York: Van Nostrand Reinhold Company.
BORGSTRÖM, BENGT-ERIK (1982). 'Power Structure and Political Speech', *Man*, n.s., 17: 313–27.
BOURDIEU, P. (1973). 'The Berber House', in Mary Douglas (ed.), *Rules and Meanings: The Anthropology of Everyday Knowledge*. Harmondsworth: Penguin Books, 98–110.
—— (1991). *Language and Symbolic Power*. Cambridge: Polity Press.
BRENNEIS, DONALD (1984). 'Straight Talk and Sweet Talk: Political Discourse in an Occasionally Egalitarian Community', in Brenneis and Myers (eds.): 69–84.
—— (1987). 'Talk and Transformation', *Man*, n.s., 22: 499–510.
—— and MYERS, FRED R. (eds.) (1984). *Dangerous Words: Language and Politics in the Pacific*. New York: Univ. Press.
BROWN, PENELOPE, and LEVINSON, STEPHEN (1976). *Universals in Language Use: Politeness Phenomena*, in Goody (ed.): 56–324.
BUNKACHŌ (ed.) (1974). *Kotoba Shirizu: Keigo*. Tokyo: Bunkachō.
BURDETT, ROSALIND (1987). *The Creative Book of Gift Wrapping*. London: Salamander Books.
CAILLET, LAURENCE (1986). 'Time in the Japanese Ritual Year', in Hendry and Webber (eds.): 31–48.
CARRIER, JAMES (1990). 'The Symbolism of Possession in Commodity Advertising', *Man*, n.s., 25: 693–706.
CASAJUS, DOMINIQUE (1985). 'Why Do the Tuareg Veil their Faces?', in Barnes *et al.* (eds.): 68–77.
CHEAL, DAVID (1988). *The Gift Economy*. London: Routledge.
CLIFFORD, JAMES, and MARCUS, GEORGE E. (1986). *Writing Culture: The Poetics and Politics of Ethnography*. Berkeley: Univ. of California Press.
COALDRAKE, WILLIAM H. (1986). 'Introduction', in Motoo.
—— (1988). 'Myths and Realities of Japanese Building', *Architecture* (Sept.): 113–17.
—— (1990). *The Way of the Carpenter*. New York: Weatherhill.
—— (forthcoming). *Raising Japanese Civilisation: Architecture and Authority*. London: Routledge.

Cobbi, Jane (1988). 'L'Obligation du cadeau au Japon', in *Lien de vie, noeud mortel: les représentations de la dette en Chine, au Japon et dans le monde indien*. Paris, Éd. de l'EHESS.

Condominas, Georges (1978). 'A Few Remarks about Thai Political Systems', in G. B. Milner (ed.), *Natural Symbols in South East Asia*. London: School of Oriental and African Studies, 105–12.

Cort, Louise Allison (1991). 'Whose Sleeves?—Gender, Class and Meaning in Japanese Dress of the Seventeenth Century', in Barnes and Eicher (eds.): 183–97.

Cunningham, Clark E. (1973). 'Order in the Atoni House', in Rodney Needham (ed.), *Right and Left: Essays on Dual Symbolic Classification*. Chicago: Univ. of Chicago Press, 204–38.

Dalby, Liza (1988). 'The Cultured Nature of Heian Colors', *Transactions of the Asiatic Society of Japan*, 4/3: 1–19.

Doi, Takeo (1986). *The Anatomy of Self: The Individual versus Society*, trans. Mark Harbison. Tokyo: Kodansha International.

Dore, R. P. (1971). *City Life in Japan*. Berkeley: Univ. of California Press.

Douglas, Mary, and Isherwood, Barron (1979). *The World of Goods*. London: Allen Lane.

du Boulay, Juliet (1976). 'Lies, Mockery and Family Integrity' in J. Peristiany (ed.), *Mediterranean Family Structures*. Cambridge: Cambridge Univ. Press, 389–406.

Dumont, Louis (1980). *Homo Hierarchicus: The Caste System and its Implications*. Chicago: Univ. of Chicago Press.

Duranti, A. (1984). '*Lāuga* and *Talanoaga*: Two Speech Genres in a Samoan Political Event', in Brenneis and Myers (eds.), 217–42.

Durkheim, Émile, and Mauss, Marcel (1970). *Primitive Classification*. London: Cohen & West, Routledge paperback edn.

Edwards, Walter (1989). *Modern Japan Through its Weddings: Gender, Person and Society in Ritual Perspective*. Stanford, Calif.: Univ. Press.

Ekiguchi, Kunio (1986). *Gift Wrapping: Creative Ideas from Japan*. Tokyo: Kodansha International.

Elias, Norbert (1978). *The Civilizing Process*. Oxford: Blackwell.

Ema Tsutomu (1971). *Kekkon no Rekishi*. Tokyo: Yūzankaku.

Evans-Pritchard, E. E. (1962). 'Sanza: A Characteristic Feature of Zande Language and Thought', in *Essays in social anthropology*. London: Faber & Faber, 204–28.

Feeley-Harnik, Gillian (1989). 'Cloth and the Creation of Ancestors in Madagascar', in Jane Schneider and Annette B. Weiner (eds.), *Cloth and Human Experience*. Washington, DC: Smithsonian Institution Press. 74–116.

Feil, Daryl K. (1982). 'From Pigs to Pearlshells: The Transformation of a New Guinea Highlands Exchange Economy', *American Ethnologist*, 9/2: 291–306.

Fenwick, Charles (1989). 'Delinquency in Contemporary Japan: Toward the Construction of a General Social Systems Theory', paper presented at the Nissan Institute for Japanese Studies, 26 May 1989.

Ferguson, Charles A. (1976). 'The Structure and Use of Politeness Formulas', *Language and Society*, 5: 137–52.

Fields, George (1983). *From Bonsai to Levis*. New York: Mentor.

Firth, Raymond (1972). 'Verbal and Bodily Rituals of Greeting and Parting', in la Fontaine (ed.), 1–38.

FIRTH, RAYMOND (1975). 'Speech-Making and Authority in Tikopia', in Bloch (ed.), 29–43.

FOSTER, JULIA (1986). *Presents*. London: Elm Tree Books.

FUJIMORI, TERUNOBU (1990). 'Traditional Houses and the Japanese View of Life', *Japan Foundation Newsletter*, 18/1: 10–15.

G'Area Communication Project (1988). *How to Giftwrap*. Tokyo: Graphic-sha.

GEERTZ, CLIFFORD (1975). *The Interpretation of Culture: Selected Essays*. London: Hutchinson.

—— (1988). *Works and Lives: The Anthropologist as Author*. Stanford: Univ. Press.

GELL, ALFRED (forthcoming). *Wrapping in Images*. Oxford: Oxford Univ. Press.

GILSENAN, MICHAEL (1976). 'Lying, Honor and Contradiction', in Kapferer (ed.): 191–219.

GOFFMAN, ERVING (1959). *The Presentation of Self in Everyday Life*. New York: Doubleday.

—— (1986). *Frame Analysis: An Essay on the Organization of Experience*. Boston: Northeastern Univ. Press.

GÖKALP, ZIYA (1968). *The Principles of Turkism*, trans. by Robert Devereux. Leiden: Brill. (1st edn. 1920.)

GOLDMAN, LAURENCE (1983). *Talk Never Dies: The Language of Huli Disputes*. London: Tavistock.

GOODY, ESTHER (1972). '"Greeting", "Begging" and the Presentation of Respect' in la Fontaine (ed.), 39–71.

—— (ed.) (1976). *Questions and Politeness*. Cambridge: Cambridge Univ. Press.

GORDON, ROBERT J. and MEGGITT, MERVYN J. (1985). *Law and Order in the New Guinea Highlands: Encounters with Enga*. Hanover, NH. Univ. Press of New England, for the University of Vermont.

GOW, IAN (1988). 'Japan', in C. Handy (ed.), *Making Managers*. London: Pitman.

GRABURN, NELSON (1983). *To Pray, to Pay and to Play*. Paris: Centre des Hautes Études Touristiques 3, no. 26.

GRAPARD, ALLAN G. (1982). 'Flying Mountains and Walkers of Emptiness: Toward a Definition of Sacred Space in Japanese Religions', *History of Religions*, 20: 195–221.

GREENBIE, BARRIE B. (1988). *Space and Spirit in Modern Japan* (New Haven, Conn.: Yale Univ. Press).

GRILLO, R. D. (1989). *Dominant Languages: Language and Hierarchy in Britain and France*. Cambridge: Cambridge Univ. Press.

GUSS, DAVID (1989). *To Weave and Sing: Art, Symbol and Narrative in the South American Rain Forest*. Berkeley: Univ. of California Press.

HEALEY, CHRISTOPHER (1990). *Maring Hunters and Traders: Production and Exchange in the Papua New Guinea Highlands*. Berkeley: Univ. of California Press.

HENDRY, JOY (1981*a*). *Marriage in Changing Japan*. London: Croom-Helm.

—— (1981*b*). '*Tomodachi-kō*: Age-mate Groups in Northern Kyushu', *Proceedings of the British Association for Japanese Studies*, 6/2: 44–56.

—— (1985). 'The Use and Abuse of Politeness Formulae', *Proceedings of the British Association for Japanese Studies*, 10: 85–91.

—— (1986). *Becoming Japanese*. Manchester: Manchester Univ. Press.

—— (1987). *Understanding Japanese Society*. London: Routledge.

—— (1988). '*Sutorenja toshite no minzokushi-gakusha—Nihon no tsutsumi bunka wo megutte*', in Teigo Yoshida and Hitoshi Miyake (eds.), *Kosumosu to Shakai*. Tokyo: Keio Tsūshin: 407–25.
—— (1990*a*). 'To Wrap or not to Wrap: Politeness and Penetration in Ethnographic Inquiry', *Man*, n.s., 24: 620–35.
—— (1990*b*). 'Humidity, Hygiene, or Ritual Care: Some Thoughts on Wrapping as a Social Phenomenon', in Eyal Ben-Ari, Brian Moeran, and James Valentine (eds.), *Unwrapping Japan*. Manchester: Manchester Univ. Press, 11–35.
—— (1990*c*). 'The Armour of Honorific Speech: Some Lateral Thinking about *Keigo*', in A. Boscaro, F. Gatti, and M. Raveri (eds.), *Rethinking Japan*. Sandgate: Japan Library, 111–17.
—— (1991). 'St. Valentine and St. Nicholas in Japan: Some Less Academic Aspects of Japanese School Life', *Japan Forum*, 3/2: 313–23.
—— (1992). 'Honorifics as Dialect: The Expression and Manipulation of Boundaries in Japanese', *Multilingua*, 11/4.
—— and WEBBER, JONATHAN (eds.) (1986). *Interpreting Japanese Society*. Journal of the Anthropological Society of Oxford, Occasional Papers, No. 5. Oxford.
HILLIER, BILL, and HANSON, JULIENNE (1984). *The Social Logic of Space*. Cambridge: Cambridge Univ. Press.
HINDS, JOHN (1975). 'Third Person Pronouns in Japanese', in Fred Peng, *Language in Japanese Society: Current Issues in Sociolinguistics*. Tokyo: Univ. Press, 129–57.
HIRAI, KIYOSHI (1973). *Feudal Architecture of Japan*. New York: Weatherhill.
HOWE, JAMES, and SHERZER, JOEL (1986). 'Friend Hairyfish and Friend Rattlesnake; or Keeping Anthropologists in their Place', *Man*, n.s., 21: 680–96.
IDE, SACHIKO (1986). 'Sex Difference and Politeness in Japanese', *International Journal of the Sociology of Language*, 58: 25–36.
II, ORION (1989). *Furoshiki Bunka no Postomodan*. Tokyo: Chūōkōronsha.
IITOYO, KIICHI (1966). '*Hōgen to Keigo*', *Kokubungaku*, 11/8: 161–6.
INOUE, FUMIO (1989). *Kotobazukai no Shinfukei: Keigo to Hōgen*. Tokyo: Akiyama Shoten.
INOUE, MITSUO (1985). *Space in Japanese Architecture*, trans. Watanabe Hiroshi. New York: Weatherhill.
IRVINE, J. (1974). 'Strategies of Status Manipulation in the Wolof Greeting', in R. Bauman and J. Sherzer (eds.), *Explorations in the Ethnography of Speaking*. Cambridge: Cambridge Univ. Press. 167–91.
—— (1979). 'Formality and Informality in Communicative Events', *American Anthropologist*, 81: 773–90.
JUGAKU, AKIKO (1979). *Nihongo to Onna*. Tokyo: Iwanami Shinsho.
KAPFERER, BRUCE (ed.) (1976). *Transaction and Meaning: Directions in the Anthropology of Exchange and Symbolic Behavior*. Philadelphia: Institute for the Study of Human Issues.
KAPLAN, DAVID E., and DUBRO, ALEC (1986). *Yakuza: The Explosive Account of Japan's Criminal Underworld*. Reading, Mass.: Addison-Wesley.
KENKYUSHA (1954). *Kenkyusha's New Japanese–English Dictionary*. Tokyo: Kenkyusha.
KOCHMAN, THOMAS (ed.) (1972). *Rappin' and Stylin' Out: Communication in Urban Black America*. Urbana, Ill.: Univ. of Illinois Press.
KOIZUMI, KAZUKO (1985). '*Noren to Kanban*', *Taiyō*, 274: 5–25.

KONDO, DORINNE (1985). 'The Way of Tea: A Symbolic Analysis'. *Man*, n.s., 20: 287–306.

—— (1990). *Crafting Selves: Power, Gender, and Discourses of Identity in a Japanese Workplace*. Chicago: Univ. of Chicago Press.

KUSAKABE, ENTA (1983). *Keigo de haji o kakenai hon*. Tokyo: Nihon Bungeisha.

KONO, T. (1984). *Strategy and Structure of Japanese Enterprise*. London: Macmillan.

KYBURZ, JOSEPH (1988). '*Engimono, Miyage, Omocha*—Three Material Manifestations of the Notion of *en*', unpublished paper presented at the 5th Triennial Conference of the European Association for Japanese Studies, Durham.

—— (1991). 'Des liens et des choses: *engimono* et *omocha*', *L'Homme: revue française d'anthropologie*, 117/31 (1): 96–121.

LA FONTAINE, JEAN (ed.) (1972). *The Interpretation of Ritual*. London: Tavistock.

LAKOFF, R. T. (1975). *Language and Woman's Place*. London: Harper & Row.

LEBRA, TAKIE SUGIYAMA (1989). 'Adoption among the Hereditary Elite of Japan: Status Preservation through Mobility', *Ethnology*, 28/3: 185–214.

—— (1990). 'The Socialization of Aristocratic Children by Commoners: Recalled Experiences of the Hereditary Elite in Modern Japan', *Cultural Anthropology*, 5/1: 78–100.

LEE, O'YOUNG (1984). *Smaller is Better: Japan's Mastery of the Miniature*, trans. R. N. Huey. Tokyo: Kodansha International.

LÉVI-STRAUSS, CLAUDE (1963). *Structural Anthropology*. Harmondsworth: Penguin Books.

MCDONAUGH, CHRISTIAN (1985). 'The Tharu House: Oppositions and Hierarchy', in Barnes *et al.* (eds.), 181–92.

—— (1987). 'The Tharu House: Deities and Ritual', in Denis Blamont and Gérard Toffin (eds.), *Architecture, milieu et société en Himalaya*. Paris: Éditions du CNRS, 261–73.

MCVEIGH, BRIAN (1991). 'Gratitude, Obedience and Humility of Heart: The Cultural Construction of Belief in a Japanese New Religion'. Doctoral thesis, Princeton University.

MAKI, FUMIHIKO (1978). '*Nihon no toshi Kūkan to "oku"*', *Sekai*, (Dec.): 146–62.

—— (1979*a*). 'Japanese City Spaces and the Concept of *oku*', *Japan Architect*, 264: 50–62.

—— (1979*b*). 'The City and Inner Space', *Japan Echo*, 6/1: 91–103.

MALINOWSKI, BRONISLAW (1972). *Argonauts of the Western Pacific*. London: Routledge & Kegan Paul.

MARCUS, G. E. and FISCHER, M. J. (1986). *Anthropology as Cultural Critique: An Experimental Moment in the Human Sciences*. Chicago: Univ. of Chicago Press.

MAUSS, MARCEL (1954). *The Gift*. Glencoe: Free Press.

MILROY, LESLEY (1987). *Language and Social Networks*. Oxford: Basil Blackwell.

MINAMI, FUJIO (1987*a*). *Keigo*. Tokyo: Iwanami Shoten.

—— (1987*b*). '*Keiihyōgen no Kōzō*', *Gengo*, 16/8: 18–25.

MOERAN, BRIAN (1984). 'One over the Seven: Sake Drinking in a Japanese Pottery Community', *Journal of the Anthropological Society of Oxford*, 15/2: 83–100.

—— (1989). *Language and Popular Culture*. Manchester: Manchester Univ. Press.

MORGAN, JAMES C., and MORGAN, J. JEFFREY (1991). *Cracking the Japanese Market*. New York: The Free Press.

MORITA, ICHIRO (1966). *Irezumi: Japanese Tattooing* (with an introductory essay by Donald Richie). Tokyo: Zuhushinsha.

MORPHY, HOWARD (1992). *Ancestral Connections: Art and an Aboriginal System of Knowledge.* Chicago: Univ. of Chicago Press.

MORSBACH, HELMUT (1977). 'The Psychological Importance of Ritualized Gift Exchange in Modern Japan', *Annals of the New York Academy of Sciences*, 293: 98–113.

MOTOO, HINAGO (1986). *Japanese Castles*, trans. and ad. William H. Coaldrake. Tokyo: Kodansha International and Shibundo.

MUNRO, NEIL GORDON (1962). *Ainu Creed and Cult.* Westport, Conn.: Greenwood Press.

MURPHY, ROBERT F. (1964). 'Social Distance and the Veil', *American Anthropologist*, 66: 1257–74.

NAKANE, CHIE (1973). *Japanese Society*. Harmondsworth: Penguin.

NELSON, ANDREW N. (1974). *Japanese–English Character Dictionary*. Rutland, Vt.: Tuttle.

NELSON, GEORGE (1967). 'Foreword', in Oka: 6–9.

NIWA, FUMIO (1966). *The Buddha Tree*, trans. Kenneth Strong. London: Peter Owen.

NUKADA, IWAO (1977). *Tsutsumi*. Tokyo: Hōsei Daigaku Shuppansha.

OGASAWARA, TADAMUNE (1985). *Zukai Ogasawara Reihō Nyūmon*. Tokyo: Chūō Bungeisha.

O'HANLON, MICHAEL (1989). *Reading the Skin: Adornment, Display and Society among the Wahgi*. London: British Museum Publications.

—— (1992). 'Unstable Images and Second Skins: Artefacts, Exegesis and Assessments in the New Guinea Highlands'. *Man*, n.s., 27

OHNUKI-TIERNEY, EMIKO (1990). 'Monkey as Metaphor? Transformation of a Polytropic Symbol in Japanese Culture', *Man*, n.s., 25: 89–107.

ŌISHI, HATSUTARŌ (1975). *Keigo*. Tokyo: Chikuma Shobo.

OKA, HIDEYUKI (1967). *How to Wrap Five Eggs: Japanese Design in Traditional Packaging.* New York: Weatherhill.

—— (1975). *How to Wrap Five More Eggs: Traditional Japanese Packaging*. New York: Weatherhill.

—— (1988). 'The Embodiment of Spirit: Reflections on Japanese Packaging Traditions', unpaginated introduction to *The Art of Japanese Packages*, catalogue for Canadian tour of an exhibition of the same name. Quebec: Musée de la Civilisation.

ONO, SOKYO (1962). *Shinto: The Kami Way*. Rutland, Vt.: Tuttle.

ORIGUCHI, SHINOBU (1945). 'Reikon no Hanashi', in *Origuchi Shinobu Zenshū*, iii: 260–74. Tokyo: Chūōkōronsha.

OUWEHAND, CORNELIUS (1964). *Namazue and their Themes*. Leiden: E. J. Brill.

Package Design in Japan (1989). Cologne: Taschen.

PARKIN, DAVID (1976). 'Exchanging Words', in Kapferer (ed.): 163–90.

PARMENTIER, RICHARD J. (1987). *The Sacred Remains: Myth, History and Polity in Belau.* Chicago: Univ. of Chicago Press.

PARRY, JONATHAN (1986). 'The Gift, the Indian Gift and the "Indian Gift"', *Man*, n.s., 21: 453–73.

PEZEU-MASABUAU, JACQUES (1981). *La Maison japonaise*. Paris: Publications Orientalistes de France.

RAHEJA, GLORIA GOODWIN (1988). *The Poison in the Gift: Ritual, Prestation and the Dominant Caste in a North Indian Village*. Chicago: Univ. of Chicago Press.

RASMUSSEN, SUSAN J. (1991). 'Veiled Self, Transparent Meanings: Tuareg Headdress as Social Expression', *Ethnology*, 30/2: 101–17.

READER, IAN (1987). 'From Asceticism to the Package Tour: The Pilgrim's Progress in Japan', *Religion*, 17: 133–48.

RICHIE, DONALD, and BURUMA, IAN (1980). *The Japanese Tattoo*. New York: Weatherhill.

RIVIÈRE, PETER (1971). 'The Political Structure of the Trio Indians as manifested in a system of Ceremonial Dialogue', in Beidelman (ed.), 293–311.

ROSALDO, MICHELLE (1973). 'I have Nothing to Hide: The Language of Ilongot Oratory', *Language in Society*, 2/2: 193–223.

RUDOFSKY, BERNARD (1972). *The Unfashionable Human Body*. London: Rupert Hart-Davis.

SAHLINS, MARSHALL (1988). 'Cosmologies of Capitalism: The Trans-Pacific Sector of "the World System"', *Proceedings of the British Academy*, 74: 1–51.

SAKURAI, TOKUTARÔ (1962). *Kōshūdan Seiritsu Katei no Kenkyū*. Tokyo: Yoshikawa Kōbunkan.

SALMOND, ANNE (1975). 'Mana Makes the Man: A Look at Maori Oratory and Politics', in Bloch (ed.): 45–63.

SANDFORD, LETTICE (1974). *Straw Work and Corn Dollies*. London: B. T. Batsford.

SEGAWA, KIYOKO (1979). '*Hachimaki*' and '*Tenugui*', in Ōtsuka Minzokugakkai (ed.), *Minzokugaku Jiten*. Tokyo: Kōbundo.

SEGAL, DANIEL A., and HANDLER, RICHARD (1989). 'Serious Play: Creative Dance and Dramatic Sensibility in Jane Austen, Ethnographer', *Man*, n.s., 24: 322–38.

SHERRY, JOHN F., and CAMARGO, EDUARDO G. (1987). '"May your Life be Marvellous": English Language Labelling and the Semiotics of Japanese Promotion', *Journal of Consumer Research*, 14: 174–88.

SHIBAMOTO, JANET (1985). *Japanese Women's Language*. London: Academic Press.

—— (1987). 'The Womanly Woman: Manipulation of Stereotypical and Nonstereotypical Features of Japanese Female Speech', in Susan Philips *et al.* (eds.), *Language, Gender and Sex in Comparative Perspective*. Cambridge: Cambridge Univ. Press, 26–49.

SINGER, KURT (1981). *Mirror, Sword and Jewel: The Geometry of Japanese Life*. Tokyo: Kodansha International.

SMITH BOWEN, ELENORE (1954). *Return to Laughter*. London: Victor Gollancz.

SPEIDEL, MANFRED, and DUFF-COOPER, ANDREW (1990). 'Sacred Places in Japan', in Werner Kreisel *et al.* (eds.), *Entwicklungstendenzen und Entwicklungsstrategien im Pazifischen Inselraum*. Aachen: Alano, 295–318.

STEINER, FRANZ (1967). *Taboo*. Harmondsworth: Pelican.

STRATHERN, ANDREW (1975). 'Veiled Speech in Mount Hagen', in Bloch (ed.), 185–231.

SWANGER, E. (1981). 'A Preliminary Examination of the *Omamori* Phenomenon', *Asian Folklore Studies*, 40/2: 237–52.

SWANSON, PAUL L. (1981). '*Shugendō* and the Yoshino-Kumano Pilgrimage', *Monumenta Nipponica*, 36: 55–84.

TAMBIAH, S. J. (1968). 'The Magical Power of Words', *Man*, n.s., 3: 175–208.

TANAKA, AKIO (1966). '*Kaisō* to *Keigo*', in *Kokubungaku*, 11/8: 155–61.

TANAKAMARU KATSUHIKO (1987). '*Mono no Shōchōsei: Iki no Minofuroshiki*', *Nihon Minzokugaku*, 171.

TRUDGILL, PETER (1972). 'Sex, Covert Prestige and Linguistic Change in the Urban British English of Norfolk', *Language in Society*, 1: 179–85.

TURNER, VICTOR (1974). *Drama, Fields and Metaphors: Symbolic Action in Human Society*. Ithaca: Cornell Univ. Press.

UNO, YOSHIKATA (1985). *Keigo o dono yō ni kangaeruka*. Tokyo: Nan' undo.

VALENTINE, JAMES (1986). 'Dance Space, Time and Organization: Aspects of Japanese Cultural Performance', in Hendry and Webber (eds.): 111–28.

VAN GENNEP, ARNOLD (1960). *The Rites of Passage*. London: Routledge & Kegan Paul.

VINING, ELIZABETH GRAY (1952). *Windows for the Crown Prince*. London: Michael Joseph.

VOGEL, EZRA (ed.) (1975). *Modern Japanese Organization and Decision Making*. Berkeley: Univ. of California Press.

WATANABE, CHIZUKO (1989). *Rappingu*. Osaka: Hoikusha.

WEINER, A. B. (1983). 'From Words to Objects to Magic: Hard Words and the Boundaries of Social Interaction', *Man*, n.s., 18: 690–709.

YAMADA, YŌKO (1988). *Watashi o tsutsumu haha naru mono*. Tokyo: Yūhikaku.

—— (1989). '*Tsutsumu*: *Nihon Bunka to Katachi*', *Shosai no Mado*, 381: 20–5.

YANAGI, SŌETSU (1972). *The Unknown Craftsman: A Japanese Insight into Beauty*, ad. by Bernard Leach. Tokyo: Kodansha International.

YANAGITA, KUNIO (1971). '*Gohei*' and '*Yorishiro*' in Ōtsuka Minzokugakkai (ed.), *Minzokugaku Jiten*. Tokyo: Kōbundo.

YOSHIDA, TETSURO (1955). *The Japanese House and Garden*, trans. from German by Marcus G. Sims. London: The Architectural Press.

ZIMMERMANN, MARK (1985). *How to Do Business with the Japanese*. New York: Random House.

INDEX